Plantations of the Low Country

Mulberry Plantation

Plantations of the Low Country

SOUTH CAROLINA 1697-1865

Photography by N. JANE ISELEY
Text by William P. Baldwin, Jr.
Researched by Agnes L. Baldwin

LEGACY PUBLICATIONS
Subsidiary of FISHER-HARRISON PUBLICATIONS INC.
Greensboro, North Carolina

Library of Congress Catalog Card Number: 84-82499

ISBN 0-933101-02-3 (Hardback Edition)
ISBN 0-933101-03-1 (Softback Edition)

Book Designer: Richard Stinely

Printed and bound in Japan

FRONT COVER PHOTO: *DRAYTON HALL AS SEEN FROM THE ASHLEY RIVER*
BACK COVER PHOTO: *DEAN HALL*

Table of Contents

Preface

ARCHITECTURE has been defined quite rightly as "the gift of one generation to the next." Barring natural or manmade disaster it is a gift that can be given over and over, a continuing legacy for all to enjoy. This is certainly true in the South Carolina Low Country. Here, in the space of a few hundred square miles, we find a unique treasure of colonial and antebellum structures: the plantation homes of our early planters, and the churches where they worshipped.

Encircled by live oaks, azaleas, and camellias, set in wide expanses of lawn, these ancient buildings have a beauty that is self-evident. Some seem familiar, a part of their neighborhood, reassuring in their permanence, while others inevitably evoke a sense of awe and mystery. Many of these buildings, especially those standing in the midst of what is today once again wilderness, may seem on first sight as mysterious a part of the past as Stonehenge or the pyramids, but the placement of a doorknob or the width of a stair reminds us that the builders were men and women no different from ourselves.

Their lives, however, were very different from ours, and therein lies a less obvious aspect of architecture's gift. We say a particular building is beautiful, that its parts are arranged into a harmonious whole, that its bays are well balanced, the columns of its porticos of pleasing proportions. In recognizing its beauty, we acknowledge that architecture is more than the four walls and roof necessary for shelter. Satisfied with an esthetic appraisal, we often forget that a building is also a portrait of the man who built it, of his family, his forefathers, and of all the residents that followed, and that collectively these biographies constitute a history. This historical legacy is especially striking on South Carolina's coastal plain, a region rich in past action and adventure, famous—even infamous—for its part in the settling and subsequent development of our country.

These plantation houses and churches are the natural embodiment of this ongoing narrative. The oldest have survived the attacks of both Indians and King, and all have stood witness to the bloody and tragic confrontation between the States. They have been the country seats of the wealthy, and shelters for the suddenly impoverished. Each one, reflecting the aspirations of its owners, and documenting their daily needs, tells what its builder wished to be as well as what he was. In each, myth is blended with reality, promise with problem, until the final story is an inseparable combination of all, and we can only look back with wonder and with admiration.

ACKNOWLEDGMENTS

To protect the privacy of the current owners of the homes included here, their names have been omitted from the text. This anonymity is certainly no reflection on their contribution, for on all sides we were greeted with enthusiastic support from Low Country residents. Their time-consuming preparation for the photography and valuable—and often unpublished—information made this project possible.

In each area there were the "movers and doers," those who run the church tours and preservation groups, and who had knowledge of the plantations in their neighborhoods. We would like to thank all these guides, especially Bennett Baxley, L. R. Connor, Mary Frampton, Helen Harvey, Mary Huguenin, Sarah Lumpkin, and Helen Maynard.

A double thanks to Gene Waddell. His knowledge of Low Country history and architecture was indispensable. His abilities as editor and diplomat brought order out of chaos. Of equal importance was Elias Bull, whose extensive research and architectural studies were invaluable. Ted Rosengarten lent a helping hand and Dale Rosengarten, Liz Seymour, Susan Walker, Sandy Wimbish, and Debra Bost all took a turn at polishing the manuscript.

Allen Stokes and Henry Fulmer helped us make use of the South Caroliniana Library's manuscript collection. Gene Waddell and his staff were equally helpful at the South Carolina Historical Society. Dr. Charles Lee and his staff assisted in identifying those houses listed on the National Historic Register and National Landmark Register. Kate Wood's Georgetown County Library was a convenient and complete source of material.

Our appreciation to Dick Stinely for designing a book of which we can be proud. In the end it was his artistic ability that molded these beautiful photographs and related commentary into its final form.

A final and very special thanks to Lillian Baldwin. She went along on the preliminary visits, assisted when the photographs were taken, and in the months since has refereed and coaxed the three of us along.

BIBLIOGRAPHY

The history of the Low Country has been documented by a number of writers who were born and raised on plantations. Often their parents had been through the War Between the States, and these writers came to their knowledge of the subject in a very personal way. Archibald Rutledge, Elizabeth Allston Pringle, Samuel Dubose, Frederick Porcher, Ann Simons Deas, Anne King Gregorie, Duncan Clinch Heyward, and others have left behind them wonderful records of this lost world. Of equal value are other family reminiscences, many of which have been published and referred to in this text.

What follows is not a complete bibliography, but simply a list of entertaining and easily available sources for those interested in the Carolina Low Country.

Dabbs, Edith M., *Sea Island Diary, A History of St. Helena Island*, Spartanburg, 1983.

Dalcho, Frederick, *The Protestant Episcopal Church in South Carolina*, Charleston, 1820.

Graydon, Nell S., *Tales of Beaufort*, Columbia, 1963; *Tales of Edisto*, Columbia, 1955.

Hilton, Mary Kendall, *Old Homes and Churches of Beaufort County, S.C.*, Columbia, 1970.

Irving, John B., M.D., *A Day on Cooper River, 1842*, Columbia, 1969.

Lachicotte, Alberta M., *Georgetown Rice Plantations*, Columbia, 1955.

Lane, Mills, *Architecture of the Old South*, Savannah, 1984.

Leiding, Harriette Kershaw, *Historic Houses of South Carolina*, Philadelphia, 1921.

McGill, Samuel D., *Narrative of Reminiscences in Williamsburg County*, Kingstree, 1952.

Rogers, George C., Jr., *The History of Georgetown County, South Carolina*, Columbia, 1970.

Smith, Alice R. Huger, *A Carolina Rice Plantation of the Fifties*, Charleston, 1970.

Stoney, Samuel Gaillard, *Plantations of the Carolina Low Country*, Charleston, 1938.

Waddell, Gene, *Charleston in 1883*, Easley, 1983.

The Historic Preservation Inventories of Buildings for the Charleston, Georgetown and Beaufort areas were most useful. These were published by the Council of Governments.

Introduction

THOUGH both the Spanish and the French attempted to settle the region of Carolina, it was the English who, in 1670, arrived to stay. Building first on the south bank of the Ashley River, they moved Charles Town a decade later to the peninsula between the Ashley and Cooper. From this harbor location a brisk trade with the mother country soon began, while the walls of the city offered relative safety from the attacks of Spanish, French, Indians, and pirates. Government was first centered here, and over the next century a rich cultural life developed, offering the new society both a winter season and a summer escape. The link between the city and the surrounding countryside remained so strong that the area has been compared to a Greek city state and its surrounding province; indeed, the high degree of cultural activity and the cultural ideals that the settlers emulated strengthens this comparison.

In these earliest years, however, survival was the first concern, and rapid profits the second. Traders traveled far inland to do business with the Indians, and planters quickly carved out their plantations along not only the Ashley and Cooper, but along other rivers and on sea islands as well. The first products that supported these settlers came from the wilderness—deerskins and maritime products such as resin—but the agriculturists of the group quickly experimented with a number of crops, and soon rice was the most successful. By 1700, the Low Country plantations were an obvious source of wealth.

The Carolina region, which had been parcelled out as rewards to supporters of Charles II, was first ruled by a government largely controlled by eight proprietors. The proprietary government, under the leadership of Lord Ashley Cooper, offered not just free or inexpensive land, but also religious freedom, and therefore attracted French Huguenots and Scotch Presbyterians as well as Swiss, German, Scotch-Irish, Welsh, and other immigrants. The predominant early settlers, however, were Barbadian planters and adventuresome English, and it was England that would set the standards for success in fashion, education, the arts, and architecture.

The earliest plantation houses could hardly have been better than the huts of the first Charles Town, but by the middle of the 1680's dwellings of consequence were being constructed along the Ashley and Cooper. By the early 1700's "the country seats of fine gentlemen" were being built in the vicinity of town, while throughout the Low Country, planters were beginning to build for permanence, comfort, and beauty.

What follows is a collection of these plantation homes and churches completed between 1697 and 1865. We have attempted here to include the most important examples that have survived on or near the coast between the Pee Dee and Savannah Rivers.

It is difficult to say how representative these houses are. The population of Colonial Carolina was small: only 3,800 white settlers are recorded in residence in 1703. The number of Low Country plantation owners was a small percentage of the total, and would remain so. In 1806, for example, only 236 whites were living on Edisto Island, and as late as 1850 only slightly over 2,000 inhabited the Georgetown District. Over the course of almost two centuries, however, it is surprising how many houses were built, especially when we consider how many were destroyed and replaced, sometimes on the same foundation.

Accidental fire was the most common cause of destruction, but war was certainly the next. While the owners could seek the protection of Charles Town in times of war, their homes were at the mercy of marauders. In 1715, the Yemassee Indians destroyed indiscriminately most, if not all, buildings that had been constructed south of Charleston. During the Revolution the British were more selective, attempting to reward the loyal with preservation of their homes while destroying those of the rebels. During the War Between the States, Union troops preserved the sea island houses taken early in the war, but Sherman and his associates burned most of what lay in their path.

Added to the depredations of man are the destructive forces of nature—hurricane, earthquake, termites, and the heat and humidity that lead to decay. Since small roof leaks eventually destroyed some buildings, we must include neglect, an unavoidable consequence of the poverty, despair, and absenteeism associated especially with Reconstruction and the Great Depression. And of course, rarest of all, some unappreciated buildings were simply torn down for scrap. In any event, what remains is as representative a sample as such a random set of experiences could produce, and though not strictly a scientific selection, together these structures and their builders give at least a rudimentary cross section of the time and place.

House Construction and Design

Although we tend to think of the Southern plantation house as the great columned mansion made famous by "Gone with the Wind" and countless book jackets, this spectacular form of Greek Revival construction was one of the least familiar to the Low Country. Only three of the sixty-six houses included here were originally built in this fashion. What we have instead is a variety of buildings built in most of the architectural styles known to America at that time. Early Colonial, Georgian, Federal, Gothic Revival, and Italianate are all represented, as well as a Jacobean castle and a French chateau.

We know of no professional architect officially connected with any of these homes (church records are more detailed), and it is possible that, except in the cases of perhaps half a dozen, none were needed. Planters sometimes designed and often contracted their houses while master builders and joiners working from plan books were capable of complex and elaborate construction.

Lumber was chosen with care and usually allowed to season. Timbers were shaped with broadaxes and adzes. Lumber was sawn by tedious man-powered pit saws. Gradually this process was mechanized as the powers of wind, water and steam were harnessed. Cypress was a favorite material, and in some instances whole houses, except their flooring, were constructed of nothing else. Resistant to rot, cypress was particularly valued for sills, siding and shingles, and its broad boards, not prone to shrinkage and easy to work, were ideal for interior paneling and carving. Heart pine was almost always used for flooring and was the usual choice for studs and other framing, especially where exposure to moisture was limited. Poplar, mahogany, and more exotic woods were usually reserved for details of trim.

Brick was readily available; the same system of shaping and kiln baking that was used in Europe was easily applied in the colony. As with cypress and pine, clay was usually a product of the site, but brick works at Medway, Boone Hall and elsewhere would eventually produce for other plantations as well. Only on the sea islands was clay difficult to procure and there oyster-shell tabby was sometimes substituted.

Masonry buildings, it should be noted, were particularly popular before the Revolution. Perhaps this reflected the old-world preference of the first generation, but with primitive milling techniques, lumber may simply have been difficult to obtain. The reason most often given for the change from masonry to wood was comfort, for brick was said to be damp and cold. Federal and Antebellum construction was usually of wood.

Roofs were sometimes made of slate or tile, but were most often covered by cypress shingles. Cypress roofs could last a builder's lifetime and could be supplemented by a second course. The obvious danger to wooden shingles was from chimney sparks, and this probably accounts for the bulk of accidental house burnings. Two of the last houses built before the War Between the States had standing-seam metal roofs, and in the years after, this favorite replacement was often laid on top of existing shingles.

Early wooden frames were joined with mortises and tenons pinned with wooden dowels. Nails were used for siding and were sometimes of local manufacture.

The most common early house plan was that of the double house which had four rooms below and four more above, with an entry into the larger of the front rooms. In later years, the entrance opened into a central stair hall. There were many variations on this basic plan, with wings and third stories common. The early Colonial cottages were small, commonly with only two or three rooms below and two loft rooms above. The basic plan of the Charleston single house or of a simple farmhouse was sometimes implemented. To these arrangements were added a variety of options, in-

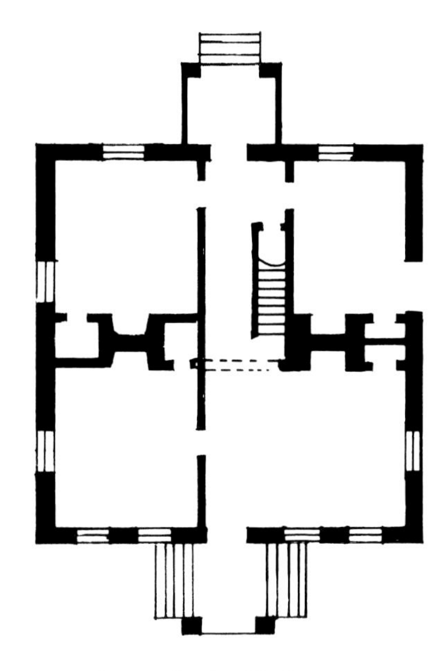

1. Early double house.

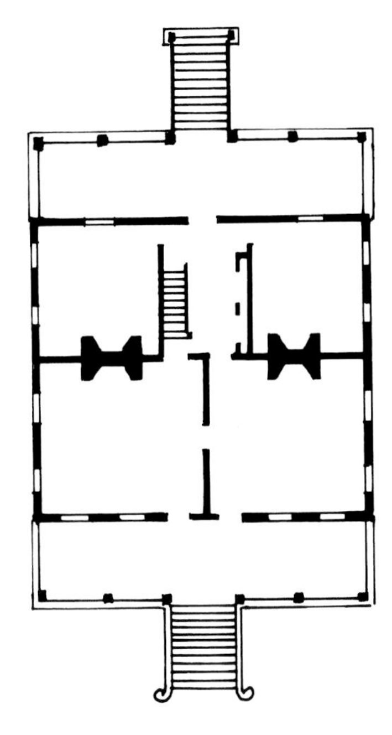

2. Double house with double entry.

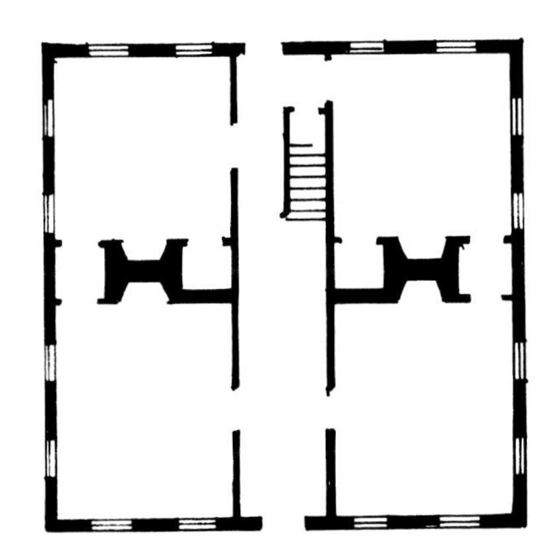

3. Double house with later entry hall.

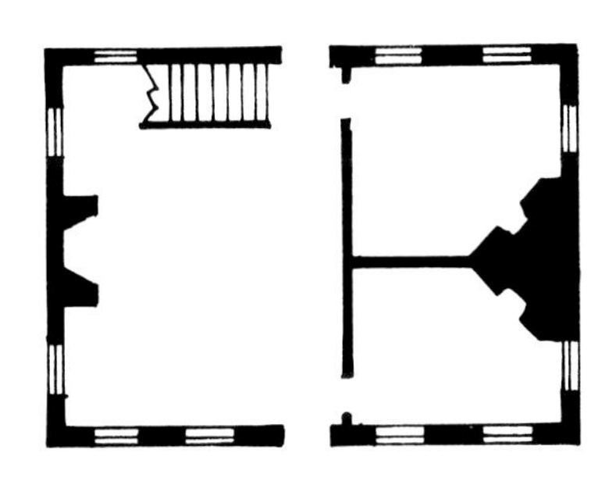

4. Colonial cottage.

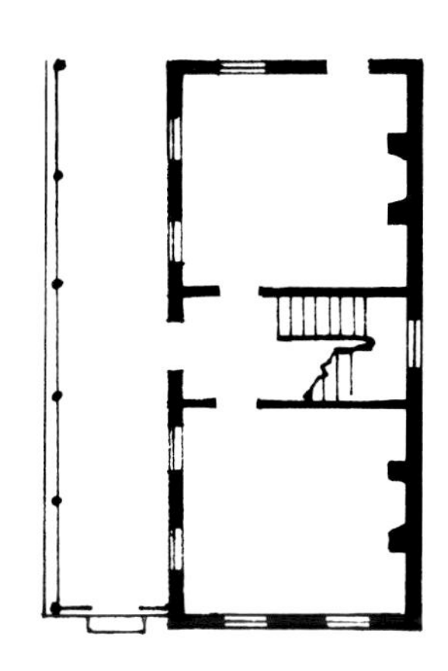

5. Single house.

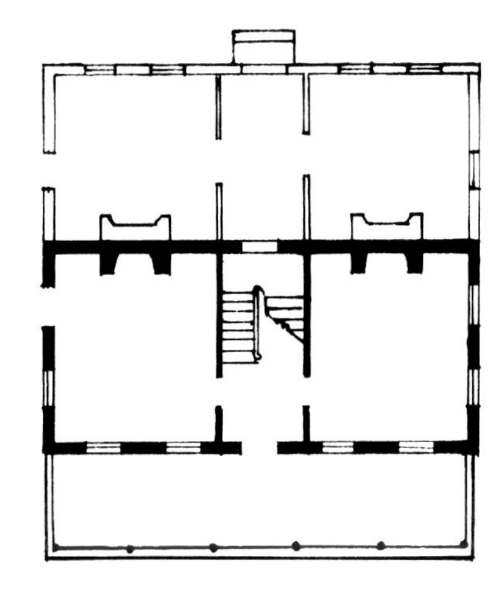

6. Farm house.

cluding raised basements, piazzas, and unusual double entries.

The treatment of interiors varied according to wealth, but generally the Pre-Revolutionary years saw extensive use of wood paneling with relatively simple mantels and trim. After Independence, wainscoting and plaster walls were common, with mantels and trim decorated with the composition ornaments made popular by the Adam brothers. By the 1820's gougework had largely replaced this imported trim, and by the 1840's most embellishment had given way to plaster walls with only simple baseboards and cornices.

Interior furnishings were as simple or as grand as the residences themselves. At one time two hundred cabinetmakers practiced their artistry in Charleston, building furniture in the tradition of Chippendale and others.

Colors—bright ones—were used inside the earliest houses, but they quickly gave way to white. Marbleizing of wood (black or gray paint was brushed with streaks of gold) and staining were popular throughout this period. Wallpaper, although not often redone in modern restorations, appeared early. Judging by Charles Fraser's sketchbook of 1800, white was the color for the exterior of frame houses, while stuccoed houses ranged from yellow to red.

The Fraser sketchbook suggests that ornamental shrubbery and oaks were not planted close by the houses, and that the most common plantings were cedars. From the earliest days of the province, numerous planters had been avid horticulturists, but during the 1800's, especially during the two decades preceding the War, a tremendous energy was put into beautifying the grounds about houses. Many of the oak avenues which we think of as timeless actually date from this period.

The People

Not every planter succeeded, for even in the best of times farming was a gambling proposition, never free from the dictates of weather, market, and sickness. Hard work, stamina and daring were the earliest attributes of successful planters—the first generations were no strangers to physical labor, and wives sometimes worked at their husbands' sides. For the following generations, it was a tireless attention to detail and a knowledge of business, agriculture, and the direction of labor that paid off.

When planters relaxed, it was to dine and drink, hunt, race horses, and dance. These same entertainments seem to have been pursued in every parish and at every income level. Southern hospitality was not a worn cliché but a necessity born of slow travel and isolated homes, and the visits of relatives and friends, especially during holidays, were often long and well celebrated. Except for entertaining at home, women could rely on little besides the church for companionship. However, their husbands enjoyed the fellowship of agriculture societies, militia, hunt clubs, and the like.

By the 1800's a certain refinement had come to the more polite and urbane elements of the plantation society, and the cavalier tradition of chivalry and romance was nurtured by the novels of Sir Walter Scott and made possible by an increasing wealth and leisure. It was then that fine wines and thoroughbred horses some enjoyed, and the ritualized conflict of lancing tournaments and duels replaced the simple fight for survival.

By the early nineteenth century, plantations were being abandoned during the warmer half of the year. Women especially enjoyed the companionship offered in summer villages, in townhouses in Charleston, Beaufort and Georgetown, or in mountain or distant spas.

Education was early appreciated by the successful planters, and before the Revolution many sons were sent back to England for an education, often completed with a study of law, medicine, or even business. During the following century, the local colleges were more fully supported, and some young men attended the best Northern schools. Throughout this period letters, diaries, and memoirs indicate a high degree of literacy, a fact also supported by the existence of numerous plantation libraries.

On the dark side was the high mortality rate suffered by the earliest settlement and persisting through the nineteenth century. Death was an everyday occurrence, and sickness was a constant problem. A tolerance for malaria was the first requirement for a successful planter, and his hope for a dynasty depended on heirs with an equally sturdy constitution. Not surprisingly, 7 of the 61 builder-planters identified here were practicing physicians. Judging by the remedies recorded in plantation log books, many others ministered to their own slaves and families.

Blamed on the bad air caused by rotting vegetation (the mysterious miasma or night vapors recorded in early accounts), malaria grew even more prevalent and deadly following the Revolution. Houses were already built on high foundations, facing into the prevailing southeast breeze, which no doubt offered some relief from both heat and mosquitoes, but the effect of the fever was now more substantial. As planters spent the months from May until November away from their country estates to escape malaria, they put less money into their plantation buildings and more into their townhouses in Charleston and elsewhere, and the architecture of the country suffered accordingly.

The Proprietary Constitution had made liberal allowances for free worship, and land grants were offered to encourage the immigration of persecuted religious groups. The governors were usually chosen from among these landgraves. Dissenters and conflict increased between this group and the Anglicans until an alliance of Barbadian planters and Huguenots won the upper hand with the passage of the Church Act of 1706. Though perhaps unfair, the Colony did profit from the passage of this act, for the Anglican church, once established as the state church, acted as a strong cultural force, providing for local education and directing charitable works. The parish system with its elected vestry strengthened early ideals of democracy, and the general religious indifference complained of by more than one Anglican priest lent itself to a healthy tolerance of other religious groups.

Another result of the act is even more concrete. Underwritten by the government and subscribed to by the wealthier planters, parish churches were well designed and solidly built. A great variety of these Colonial structures still grace the parishes.

Other faiths were equally active, but lacking government backing and not being inclined to worldly display, they built churches that were less likely to survive. However, both Baptists and Presbyterians are represented here with buildings which are no less impressive.

Politics and War

The term Low Country is so often associated simply with the geography of sea islands, marshes, and swamps lying at sea level or only slightly higher that we tend to forget the political connotation of the words. New arrivals moving beyond the coast soon outnumbered the established planters, and after the Revolution it was necessary to share political power with this newly settled "upper district." The "lower district" or Low Country extended then little beyond the limits of the rice-growing portions of the rivers, but with the spread of cotton cultivation, the boundaries of the slave-owning oligarchy were pushed further inland toward the fall line.

John Locke's constitution for the proprietary government set up a land-based aristocracy of landgraves and caciques, but except for these titles, it was largely ignored. The colonists formed their own aristocracy on the basis of achievement—an achievement that was most often measured by the acquisition of land. Even after the Revolution, this principle of representation on the basis of land and wealth continued in force, finding its way not only into the state's constitution, but into that of the Federal government as well.

It is not surprising, then, to find that of the 61 identified house builders in our survey, 16 had served as Governor, Lt. Governor or in the Colonial or state assemblies, and that many others held positions of power such as that of Colonial treasurer, ports collector, provost marshal or intendant. Early political leadership came almost solely from the planter class, and in times of constantly shifting alliances, conflicts, and rebellions, public service was an inescapable privilege.

The same was true of military service, and the colonists, continually under attack during the early days, formed a habit of military self-sufficiency that stood them in good stead during the Revolution. Neutrality was impossible, and most of those capable rode with Marion or served under other partisan generals. Several of the house sites presented herein were battlefields. Such was the nature of this war, however, that five residents (usually old and more established planters) out of the 18 residents of Pre-Revolutionary houses represented were loyalists or swore allegiance to the King and suffered for their mistake. Much less division existed during the next great conflict. Most Low Country planters were ardent secessionists, and almost all who joined in the Confederate cause met with more or less equal ruin.

Agriculture

We think so often in terms of Southern or Low Country plantations that we often forget that the plantation was not a local or regional phenomenon but a basic unit of agricultural production in many of the tropical and subtropical regions of the world—sugar plantations in Cuba, rubber plantations in Indonesia. Wherever colonial governments extended, they encouraged large holdings worked with gang labor (often slaves) and production of a single crop which was profitable for both the grower and the mother country to which it was shipped.

It was almost inevitable, then, that the Barbadian planters would bring this system with them to Carolina and quickly add rice, and later indigo and cotton, to the colonial roster. It was tragic, nonetheless, for what makes the South—and most especially the Low Country—unique was the long-continued profitability of this system and the fact that it took a war to end it. Both rice and Sea Island cotton were what we call today "labor intensive" crops, and from the earliest days profits were converted into slaves and slaves into profits in a vicious cycle that could only end in disaster.

The baronies of the proprietary period were quickly broken into more manageable allotments, so that Low Country plantations were by no means the endless tracts sometimes envisioned. "A little plantation is a sorry undertaking," wrote James Petigru of his own 200 acres in rice, but some of our planters managed well with holdings no bigger, and the plantings of 1,000 acres or more (the combinations of several plantations) were unusual. These figures, of course, do not include the high land holdings, which like the house, were sometimes included in land transfer gratis. Higher ground could be used to provision the inhabitants, but a great amount was not needed.

Introduced in 1672, rice was first planted in cleared swamplands bordering rivers, and reservoirs were sometimes built at headwaters. This land was eventually exhausted, however, and in the mid 1750's planters found that they could get better results by diking marsh flats and controlling fresh water that was needed for cultivation by a system of trunks.

In the 1740's, the War of Jenkins Ear between England and Spain cut the colony off from its rice market. Eliza Lucas and others began experimenting with indigo, which took much skill to grow and even more care to process, but the rewards were great. This became a principal moneymaker, especially for the more upland properties, until the coming of the Revolution.

Sea Island cotton was first grown in South Carolina in 1791, and within a decade it had transformed the economy of the coast. With a fiber twice the length of upland cotton and a price as much as six times higher, this was another crop that could be produced successfully on small acreage. Three hundred and fifty acres was the average size of a sea island plantation, and rarely did one exceed 1,000 acres. Much labor was nonetheless needed, for the long-staple crop required numerous hoeings and intensive fertilizing with manure and marsh muds, and the preparation for English market was time consuming.

This same seed sown on the mainland just in from the islands produced a slightly less valuable "main" or Santee cotton. (It could not be planted far from the coast.) In the interior, the crop was upland or short-staple cotton that required less labor and could be grown on greater acreage. Still, judging by the plantations we reviewed, 2,000 acres of upland cotton was considered a large undertaking.

Rivers

Rivers were the first routes for travel by the settlers and continued to serve them until the War Between the States. It was therefore natural for the earliest planters to build along their banks (bluff sites where possible) especially in the coastal rice-growing regions. Early canoes, cut from cypress logs, and large plantation flats made the journey along with sloops and small schooners which were later joined by steamboats.

Roads, which were built not long afterwards, often ran parallel to the rivers. Thus, many houses facing the Santee, Cooper or Ashley Rivers had river-road entrances as well. These roads were notoriously bad, often impassible, and could never be counted on for the transport of crops to market.

We have organized our discussion of the buildings in this book around the Low Country rivers, for they were, after all, linear neighborhoods. Along their banks could be found relatives, friends, and allies. Planters would grow the same crop in similar fashion, share similar concerns and joys, and most important, their houses would, as Mills Lane points out, be architecturally linked by this "localism." The exceptions to our plan are the sea islands of St. Helena and Edisto, and the remote communities of Eutawville and Black Mingo which offered a more conventional clustering of houses. Influence, it should be noted, was not always based on proximity, for relatives separated by 100 miles were still arbiters of design and taste. A Charleston townhouse like Miles Brewton's might be copied both north and south of the city, and workmen using a plan book could produce similar results in more than one location.

The rice rivers were the Waccamaw, Pee Dee, Black, Sampit, Santee, Cooper, Ashley, Edisto, Ashepoo, Combahee, and Savannah. Except in one case, houses along the last three and the Sampit have not been included in this book because of the ravages of war and ordinary bad luck. The first three rivers have been grouped along with the inland Black Mingo community. The Santee is discussed both at its delta and inland at the Eutawville community. The Ashley and the Cooper are reviewed together along with the nearby Christ Church Parish, Kiawah, and Johns Island. Edisto Island and the Edisto River are bracketed, and the final section is St. Helena Island and its accompanying area of Port Royal Sound.

Within these neighborhoods the houses are arranged chronologically, and when possible, the exact date of construction and the builder's name are given. Each question we answered invariably suggested another to be asked, a process which leads naturally beyond the confines of this book. But here we hope is an enjoyable beginning.

This earliest of South Carolina's houses is distinguished by its simplicity and the accommodation it makes to the climate. This chimney heated the original four rooms, and the front and back piazzas provided summer shade for Huguenot Simons and his many descendants.

Ashley and Cooper Rivers System

WITH Charleston built on the peninsula between the Ashley and Cooper rivers, it was inevitable that the earliest trade and settlement would occur along their banks. The Ashley offered easy access to the city and was soon a favorite site of the elegant country seats. Of these, however, only Drayton Hall remains. On the Cooper more extensive rice production was possible, and a great variety of houses have survived. In the early days land which was closer to the seacoast was less heavily settled, and the homes were usually modest.

In 1842 Dr. Irving offered an account of this region in his *A Day on Cooper River* which Louise Stoney annotated almost a century later. Here's her description of Dr. Irving's world:

> As we travel with him, imagination must show us his plantations at their best, the fields covered with the rich green of the growing crop, or the gold of the harvest, checked here and there with the mauves and purples of the canals and ditches and the darker green of the cedar-sentinelled banks. Among them, gangs of Negroes are working their tasks; or later in the season, are shaking rattles and cracking whips to "mind off" the rice-birds that would gorge on the ripening grain, perhaps shooting them, for rice-birds were always a harvest delicacy; or later still, cutting or shocking the rice or loading it into waiting lighters to be carried to the rick yards.

Middleburg

Built ca. 1697, Cooper River, St. Thomas and St. Dennis' Parish, Berkeley County, National Register, National Historic Landmark.

The oldest part of Middleburg was built about 1697 by the French Huguenot Benjamin Simons; the house takes its name from the ancient capital of the Zeeland Province in Holland. There is little here to indicate these cosmopolitan origins, however, for what is exceptional about this earliest of all South Carolina houses is its simplicity and the accommodation it makes to the local environment.

Extensive and careful restoration currently underway suggests that the original house was relatively small. Two single rooms below and two more above were heated by a single, central chimney. Wide piazzas provided shade across the front and back of the house.

Records show that the fifth of the Simons' 14 children was born in the newly completed house, and that the second, Benjamin, subsequently raised a large family there. The next generation included Catherine Chicken whose childhood ordeal in the Strawberry Chapel graveyard has become legend, and Lydia Lucas whose husband and famous father-in-law, Jonathan Lucas, built the first toll rice mill on the grounds. Lucas' invention did for rice cultivation what the cotton gin did for cotton. Remains of this mill still exist, as well as an interesting nineteenth century commissary building, where slaves were once jailed, and a carriage house.

The additions, outbuildings, mill, and a symmetrical formal garden with camellia alleé and reflecting pond, all suggest that from relatively humble beginnings the family quickly prospered, beautifying and enlarging their residence accordingly. This process was not as unusual as the fact that the plantation remained in their hands until 1981.

Much has been made of the Huguenots' contribution to South Carolina, a contribution that seem's on first glance to far outweigh their numerical presence. Early settlers like the Simons, forced from their homeland by religious persecution, brought to the frontier attributes that were sometimes less evident in their freebooting speculator neighbors—hard work, thrift, piety, and a general temperance of thought and action. Middleburg still stands as a testament to these early Huguenots' virtues.

Medway Plantation

Built ca. 1705, Back River (Medway River), St. James Goose Creek Parish, Berkeley County, National Register.

Hidden at the center of at least one great addition to Medway is a small European-style dwelling built by Elizabeth and Edward Hyrne in 1705. Though the plantation is still probably the oldest brick house in South Carolina, it was not, as previously thought, built by the Dutch settler John D'Arsen in 1686. Extensive correspondence and the recent discovery of four bricks embossed with a coat of arms indicate that this first house which is usually associated with the second husband of D'Arsen's widow, Landgrave Thomas Smith, burned in 1704 and was replaced by the Hyrnes. If original to the present building, the crow-stepped gables long credited to the Dutchman may have been added because they were also popular in Elizabeth's native Lincolnshire.

We should not be disappointed by this revelation of new builders, however, for although the turbulent history of the anti-Anglican Landgrave is entertaining, there is romance enough in the story of the young English couple. Edward Hyrne was a Norfolk merchant fleeing from his creditors and from charges that he had misapplied £1300 of government money. His 18-year-old wife, who was the daughter of a baronet, was waiting for her inheritance to pay for the new plantation. They wrote home glowing accounts of their prospects:

Two large additions swallow up the original Medway, but the result is a harmonious whole, drawn together by stucco, fig vine, and a unity of design. In spite of much research, the origins of the distinctive "Dutch" gable-ends remain a mystery.

This well-lit dining room with its high ceilings was added in 1855. Furnishings are eighteenth-century English. The portraits of family members were painted by English artist Simon Eloise.

Already possessing 150 cattle and "the best Brick-house in all the country," they needed only slaves to produce "rice, pitch, tar, cedar, cypress, oak and other timber." Tragedy followed. The house burned, their infant son died, their only slave was killed by a rattlesnake, and the inheritance was withheld. They built the present house without first obtaining clear title to the land. Edward was jailed in England, even as in the colonies he was made port inspector. When the inherited money arrived, it was apparently too late, for the property had reverted to Thomas Smith, II, who then married a daughter of Hyrne's first marriage. "Living was rapid but there were bonuses for the astute," wrote Mrs. Stoney of Landgrave Smith, who lies buried on the grounds.

The Hyrnes' residence, as advertised in 1738, appears as "a good brick house 36 feet in length, 26 feet in breadth, cellars and kitchen under the house." It can be identified today by a poor quality brick laid in English bond. This dwelling was originally a story and a half. Divided into three or possibly only two rooms, the main floor had front and back entrances and fireplaces at each end. The ceilings were low and the windows relatively small.

The west wing is known to have been added in 1855, and since the opposite wing has the same common bond for brick work and identical chimney arches, it was probably built at the same time. The oldest section gained a story then, if not before. Incorporating some of the old beams, the roof is largely nineteenth-century work. Drawn

During a 1930's restoration the cypress interior of the living room was transferred from nearby Pine Grove. The "plantation desk" in the far corner is now being reproduced by the Historic Charleston Foundation.

together by stucco, fig vine, and a unity of design, the result is a harmonious whole that both celebrates and belies its humble beginnings.

Though numerous owners held the property during the prosperous period of rice growing, the last one bears special mention. The Stoney family purchased the Medway or Back River Plantation in 1835, and it was they who made the additions, planted the double oak avenue, maintained the gardens, and adopted the name of Medway, either from their own creek or England's Medway River.

Samuel G. Stoney, who grew up here, was the grand story teller and historian of the Low Country. He and his brother-in-law, Albert Simmons, and Samuel Lapham prepared the inimitable *Plantations of the Carolina Lowcountry,* a book that did much to celebrate and preserve what we see today.

St. Andrew's Parish Church

Built ca. 1706, Ashley River, St. Andrew's Parish, Charleston County, National Register.

At the end of "stormy debate" in 1706, Governor Nathaniel Johnson requested that the Commons House do its duty "Both to God and the Public," and establish a church. The Assembly replied that they would take this opportunity that "God has given us of Fencing our Vineyard, and making the Hedge about it as Strong as we can." The resulting "Church Act" of that year established the Church of England as the official faith of the colony, with the cost of building churches and parsonages to come from a tax on skins and furs and the contributions of individuals. Ten parishes were laid out at that time to serve not only as ecclesiastical boundaries, but temporal as well. There were no courthouses in the province, so the church doubled as an administrative center, and the vestrymen and church wardens elected by their parishioners served as public officials. St. Andrew's Church was the oldest of these authorized buildings in the state.

Over the west door a tile proclaims: "SUPERVI 1706, J. F.-T. R." Jonathan Fitch and Thomas Rose, church wardens and bricklayers by trade, had supervised the construction of the original 40′ × 25′ building. In 1723, £900 was authorized for enlarging the church, and the addition of transepts and a chancel created a cruciform layout. In 1764, a fire partially destroyed the structure; when it was repaired the window behind the altar was closed in. At this time wardens began to rent or sell pews to raise money to repair and maintain the building.

Though originally one of the wealthiest parishes in the colony, its economy depended heavily on indigo. St. Andrew's planters were doubly hurt in the years following the Revolution by the cessation of the indigo bounty, through which the Crown had supported the indigo trade, and the disenfranchisement of the Anglicans. This church, like several in other parishes, fell into disrepair. In 1855, however, vestryman William Izard Bull undertook an extensive renovation and a dated sketch of his plan was left behind on the wall. When the faithful and long-serving Reverend Grimke-Drayton died in 1891, there were few communicants remaining in the parish, and the church closed its doors. In 1933, The Colonial Dames made some repairs, but not until 1948 was the church opened again for regular services.

The crucifix layout, brought about by a 1723 addition, is rare in colonial America and unique to our survey. Viewed from the gallery, the elliptical vaulting of the ceiling merges above. Below is a baptismal font supported by the Pelican symbol of the Society for the Propagation of the Gospel in Foreign Parts. The altar rail is supported by a unique cast-iron balustrade, and the towering reredos with the creed, Lord's Prayer and Ten Commandments probably date from 1723.

Springtime wisteria and dogwood frame the Rev. Le Jau's St. James Goose Creek Church. Above the door are the flaming hearts of charity and the pelican symbol of the Society for the Propagation of the Gospel in Foreign Parts.

St. James Goose Creek Church

Built ca. 1708, St. James Goose Creek Parish, Berkeley County, National Register, National Historic Landmark.

In *Early English Churches in America,* St. James Goose Creek Church is described as "one of the best preserved and most interesting churches in the United States." The author of this work attributes the inspiration for the church to the Barbadian planters who attended it, claiming "its rectangular mass, elaborate stucco ornament, yellow walls, and jerkin-head roof bear an unmistakable relationship to the West Indian architecture." Sam Stoney fails to mention this debt, but certainly corroborates the grandness of the building, calling it "the most baroque piece of decoration that I can find any record of in any of the English churches of the 13 colonies." The colored high relief of the dadoes and the unique royal coat of arms certainly substantiate this latter claim. The coat of arms is of particular interest, because it was on its account that the British spared St. James while destroying or vandalizing most other parish churches.

Outside the building on the pediment above the entrance is another feature that bears mention, a pelican tearing at her breast to feed her young—the symbol of the Society for The Propagation of the Gospel in Foreign Parts. From its beginning the church had been serviced by this society and a member of the society. Dr. Francis Le Jau, was the second rector of the parish. The symbol possibly had another meaning for the good doctor, for in 1707 he wrote:

> I thought all the great noise . . . was grounded upon true zeal for the Glory of God . . . But I assure you it is far from it, revenge, self-interest, engrossing of trade, places of any profit and things are ye motives that give a turn to our affairs.

Both parties of the religious conflict between Anglicans and Dissenters were the objects of his castigations, but the Barbadians of his own parish seemed the most guilty. Known collectively as "the Goose Creek men," they were the bane not only of Dissenters but of some governors of the colony.

Le Jau's relationship with his congregation can only be guessed at by reading between the lines of his letters to his England-based superiors. He spoke out against cruelty to slaves and strongly advocated the religious instruction of both Blacks and Indians. His readiness "to call evil, evil" no doubt made him enemies, while his strict construction of the Scriptures—he refused to marry a man to his widowed sister-in-law or baptize an infant whose parents weren't baptized—must have been exasperating to his frontier parishioners. At the high point of his tenure he had built church attendance to a very respectable 100 but modified his methods slightly: "I see that if I should be too earnest in shewing the Evil and opposing it, it would be worse. I am satisfied and bless God for the good disposition of some men who are sincerely believers and Religious, the number is not great."

St. James Goose Creek is considered by many to be the most interesting and beautiful of rural Anglican churches. The enormous reredos seen here is made of painted plaster. High above the pulpit is a decoration that cannot be found in any other colonial church—a royal coat of arms.

The Yemassee war of 1715 struck hard at the neighborhood, greatly reducing the size of the congregation, and, with his health failing, Le Jau's ministry never recovered. His parsonage took a long time to build, and from its beginning his pay came seldom and was sometimes in depreciated paper money. Worst of all, the beautiful new church, begun at his arrival and completed at his parishioners' expense and convenience, appears not to have been occupied until shortly after his death.

One last letter details his final situation:

> God's will be done, I am Resigned to it by his Grace, and tho I have cruelly suffered by Sickness & want, & my family has no cloathes these 2 years & I lost 2 young Slaves that dyed, and a third is adying I fear, and I am above 200 £ in debt for bare Necessaryes & we live very hard upon Indian corn we buy at 10 sh a bushell with little or no meat. Yet trusting in God & depending upon my Superiors bounty & favour I am satisfied to Serve while I live here If my hon'd Superiors please to command me.

So ended "good Dr. Le Jau," leaving behind an impoverished widow and family, and in his lengthy correspondence a vivid record of life in the early colony. He was buried beneath the altar of St. James Goose Creek.

Mulberry

Built ca. 1711, Cooper River (West Branch), St. John's Parish Berkeley, Berkeley County, National Register, National Historic Landmark.

With the building of Mulberry in 1711, Low Country architecture took a giant step forward, or perhaps we should say backwards, for the Jacobean building is clearly a transplant from the English countryside, heralding the inclination to impose the finished product of old-world culture on the raw frontier. Just how raw that frontier was is illuminated by the fact that the high bluff on which Mulberry sits had been an Indian camp only a few years before. "The castle," as it was called, gave shelter to the refugees of the Yemassee Indian War shortly after its construction. Although Stoney, Simons and Lapham suggest that the four flanking

Thomas Broughton built here on the frontier's edge in 1711. The four small flankers with bell-shaped roof and the jerkin-headed roof of the main house suggest a truncated version of a Jacobean castle.

towers were intended only for decoration, the structures were said to have contained trap doors that led to powder magazines. In addition, two small cannons were reportedly dug up on the property in 1835.

It is difficult to resist speculating that the classic fortification layout of the house had a utilitarian appeal to builder Thomas Broughton. Though regarded with affection by his wife and having the rare approval of minister Le Jau, this castle builder was far from timid, and clearly considered action the better part of valor. A prosperous planter and politician, he was a soldier and Indian trader as well, and there is much in his conflict-studded biography to suggest that he would have expected his magnificent mansion to serve on occasion as a fortress.

The original grant of the property was to Sir Peter Colleton, and his son swapped it with Broughton, who had built his home there before making a proper survey of the dividing line. Apparently the house builder felt the same proprietorship toward political office, for when an opponent's bribe cost him the governorship, he marched on Charleston with an armed group demanding the position. He was not successful, but did serve as lt. governor much later, governing the province with a singular generosity to himself until his death in 1737.

As to the house itself, there is much to be admired. Careful and thoughtful restoration has not diminished the original. It is considered most unusual for the jerkin-head gables on its gambrel roof and the four small flankers which tend to disguise the conventional floor plan. The entrance leads into the larger of two front rooms. A stair hall at the rear gives access to the back rooms and bedrooms above. This arrangement was the most common one until the introduction of the central hall in the middle of the eighteenth century. While the present interior shows the influence of Post-Revolutionary prosperity and taste, much of the original molding and panels can still be seen in odd places. On the upper-floor doors and keyed mantels the molding and panels have remained intact.

Laid in English bond, the brickwork is exceptional. The watertable is of molded bricks, and above this the dappled effect of randomly placed glazers is accented with solid red bricks at the corners and above the windows.

The four weathervanes at the corners of the house are capped by royal crowns, and above the entrance porch is a woodcarving of a fruited mulberry twig encircled by a horseshoe. The woodcarving, no doubt, is a tribute to the large mulberry tree that had been growing wild on the site, and the luck that accompanies iron-willed perseverance.

A mantel from the Federal period has replaced the earlier bolection molding. In the far corner, a recessed shell wall cabinet holds a blue and gold Rockingham tea set. The entry into a front parlor with the stairwell at the rear was the most common floor plan of early South Carolina.

When Col. William Rhett built his house, no other houses stood between his house and his Cooper River wharf. It was from that wharf that he sailed to bring back the notorious pirate Stede Bonnet.

William Rhett House

Built ca. 1712, Cooper River, St. Philip's Parish, Charleston County, National Register.

It is hard to imagine that a dwelling so engulfed by urban congestion was once the main house of a plantation whose southernmost boundary was the limit of Charles Town. Such is the case, however, with the William Rhett House. The second mansion to exist on the site, it was built about 1712 by the well known Col. William Rhett.

In considering the builder's career it should be remembered that in Colonial Carolina the line between pirate, privateer, and private citizen was sometimes thinly drawn. Wars with France and Spain were a good excuse for licensed privateers to prey on enemy ships, and Charles Town welcomed the booty this trade provided. The colonists were accused, and perhaps rightly so, of dealing with anyone—Spanish, French, Indian, or pirate—who didn't threaten their immediate existence. Yet by 1718 matters had gotten out of hand. The notorious Blackbeard blockaded the port and made demands on the captive citizenry. The time had come to address the problem, and Col. Rhett set out from here in search of brigands.

His capture of the gentleman pirate Stede Bonnet, the trial, escape, and subsequent hanging of Bonnet are all familiar elements in the Low Country pageant and typify Col. Rhett's adventuresome life. As historian Wallace reminds us:

> There are few men in South Carolina history of such contradictory character as this greedy, violent, vulgar, lawless, brave, impulsive, generous, loyal churchman and pirate fighter. Greedily violating law and propriety for bigger profits, insulting the noble and courteous Governor Craven too vulgarly for quotation, trying to kill Governor Daniels in a quarrel over authority, fighting Stede Bonnet to the finish, magnanimously offering him mercy and refusing before the court to share in the prize money, he represents the raw material of violent passions, powerful personality, and untamed willfulness.

Rhett's mansion seems to have sheltered more than its share of strong-willed personalities. His widow was a merchant and a controversial figure. Medway's Elizabeth Hyrne makes much of her generosity, a calculated trait that entered Mrs. Rhett into several lawsuits with friends she was said to have swindled. She remarried one of her husband's old allies, Chief Justice Nicholas Trott, "the hanging judge" who had guided the province through the turbulent years of church conflict, Indian wars, widespread privateering, and the removal of proprietary government. By the next century the house had passed to Col. Wade Hampton, and in 1818 Wade Hampton III, Civil War general and Reconstruction leader was born there.

The house is similar to other early Charleston houses, square on a raised foundation. Today, closed piazzas that date from about 1800 conceal both the front and rear entrances. A 1739 depiction of the area shows the house in the midst of a park, approached from both directions by avenues that in the present day would lead to King and East Bay Streets.

Hanover

Built ca. 1716, St. John's Parish Berkeley, Berkeley County (Moved to Clemson University Campus), National Register, National Historic Landmark, Open to the Public.

"Peu a Peu" reads the inscription on the top of Hanover's large chimney, the opening words, we believe, of an expression still common in France today—"The bird builds his nest little by little." When the Huguenot Paul de St. Julien built here in 1716 this was clearly his guiding principle, for the house, though small, was built with great care.

Much is known about the construction of Hanover. The building lay within the bounds of the area to be flooded by the Santee-Cooper project, and being designated by architectural historian Thomas T. Waterman as the most architecturally significant structure in the affected area, it was

Relocated on the campus of Clemson University, gambrel-roofed Hanover is a clear expression of old-world design. Enough brick went into the large basement to "hold a small Eiffel tower," family records recall. This foundation and the distinctive construction of the flues are attributed to the gallic origins of the Huguenot builder.

dismantled by Clemson University and reassembled on its upcountry campus. In the squarish dormers and various elements of simple trim, Waterman discovered tiny clues of a French influence unique to the colonies, and so saved it from the rising waters. The clapboard siding and gambrel roof are typical of much of America's early architecture and would not look out of place in either New England or Williamsburg. Sadly, 20 or more other houses dating from the early 1800's were not similarly graced by this Gallic touch and, despite their richly carved interiors, were inundated.

Nevertheless we can be happy that we have Hanover. Every plank of it, from floor joist to roof, was sawn from cypress, worked by hand and fitted with exceeding care. As at Middleburg, the interior walls are of battened board, but here they were trimmed to simulate paneling. Most unusual is the ceiling of the gable rooms. The wood paneling was applied in a reversed ship lap so that any rain water which passed through the shingles would be shed from the interior.

Piazzas on both the front and back of the house were not restored. It was decided, after consulting Samuel G. Stoney, that the original structure would have stayed closer to the European ideal and made little concession to our climate.

This ideal is certainly respected in the brick-

Over the parlor fireplace is a portrait of early owner René Ravenel. Beside the hearth are dummy board figures or silent companions, a popular decoration of the day. With an emphasis on instruction as well as handsome decoration, the entire house has been beautifully furnished by the Spartanburg Committee of the National Society of Colonial Dames and is open to the public.

work, which is always referred to as the house's most Gallic element. Rare for the Low Country, the march of the flues to the peak of the roof can be followed easily. The family records describe these as "being really two chimneys at each end of the house, one built outside the other from the ground to the top." The records comment as well on what would have been an equally French construction. Left behind beneath the surface of the lake was evidence of a terrace and a large basement that included kitchen and cooling rooms. Enough brick went into its foundation, it is said, to "hold a small Eiffel tower," and the story is told that the builder's original plans to build a solid brick house were abandoned when he had used three kilns worth of masonry on foundation and chimney alone.

As for the family that lived there, eldest daughter Mary married Henry Ravenel, son of another Huguenot arrival, and for two centuries their descendants kept possession of the house. Within the brief space of a few pages of reminiscence and genealogy, the *Ravenel Record* gives a vivid picture of Hanover life. Birth, courtship, marriage, death, war, storm, farming, horseraces, and deer hunts all get passing mention. In 1752, a hurricane blew down many of the outbuildings. Stephen Ravenel appears only incidentally as South Carolina's Secretary of State and is remembered as "devoted to hunting and killed a great many deer." Large for his age, Paul Ravenel was captured by the British when he was a child, but was released when found playing marbles with a Negro boy. On reaching the age of fourteen he served briefly with Marion's Men. The residents of Hanover never appeared to have prospered grandly, but neither was life dull.

Johns Island Presbyterian Church

Built ca. 1719, St. John's Parish Colleton, Charleston County, National Register.

At the end of the 1600's Reverend Archibald Stobo was returning to Scotland from the failed colony of Darien at the Isthmus of Panama. As his ship lay off Charles Town he was invited ashore to conduct a service and brought with him his wife and a dozen others. A storm struck, sinking the ship with all aboard, and so the pastor found himself serving the city's Independent Church. Seeing that the outlying area was not being reached by the Presbyterian faith, he organized five churches encompassing Johns Island, James Island, Edisto, Charleston, and Willtown. He preferred these rural churches to city congregations and preached regularly at the church he established on Johns Island.

Often church members left wills providing for the clergy. In 1740, one specifically indicated that the minister "conforms to the Kirks of Scotland," hinting at least that this congregation or its ministers had expressed conflicting ideas. Rev. George Whitefield preached here in that year, while he was being tried by the Bishop's Commissary in Charles Town. A leader of the Great Awakening, the religious revival that swept the colonies in the mid-1700's, Whitefield had strayed from the prescribed service of the Book of Common Prayer and was preaching a concept of human responsibility and free will. His appearance at the Johns Island church suggests that the congregation was affected by this revival which would lead eventually to the founding of the Methodist church.

Also of national importance was the resident minister Moses Waddell, who in the tradition of John Knox had established a famous Presbyterian academy. Education was of the utmost importance to these church members, as anyone who has attempted to follow an ancient Presbyterian sermon can attest. A church-sponsored education could also encompass the secular uses of planting, commerce, and law. A list of church-related pamphlets indicates that the well-informed congregation debated nonreligious crises with fervor, and John Townsend of Bleak Hall and others appeared to have argued at length for secession.

The structure is the oldest Presbyterian frame church in America. Original to the building is the 35' × 38' portion that symmetrically includes two entry doors, two side doors, and two windows. This portion is thought to date from about 1719, but did not take on its present appearance until a 1792 remodeling. Another extensive alteration in 1823 moved the far wall back 20 feet, adding a new rear façade with two palladian windows and accompanying doors with semi-circular fanlights of palm-like wood construction. As we will see later at Edisto, the gallery of the interior is lit by a set of smaller, rectangular windows above the fanned ones. Here, however, there is no portico or cupola to alter the chaste meetinghouse lines.

Though not Anglican, the Johns Island church followed some general characteristics that Stoney mentioned for the early parish churches. Because of the widely distributed congregation, the building was located on both a main road and near a river. There was no provision made for a bell tower that would be useless in the country. Though not enjoying the privileged status of the Anglicans, the Presbyterian churches south of Charleston seemed to serve much the same functions: dispensing charity, educating, acting as a public forum and, as the Huguenots did in the north, instilling in the community the virtues of hard work, thrift, and moderation.

A large Sunday school addition across the rear tends to disguise the modest scale and classic meetinghouse lines of one of America's oldest Presbyterian churches.

The small symmetrical building with jerkin-headed roof was strategically located to serve the congregation near the T of the Cooper River. In the foreground is the Ball family enclosure.

Strawberry Chapel

Built ca. 1725, West Branch of Cooper River, St. John's Parish, Berkeley County, National Register.

In 1718, James Child bequeathed an acre and a half of land as a burying place for "the inhabitants of Childsbury Town and all within the western and eastern branches of Cooper River, as it is bounded out with locust trees and cedar fence." He reserved for his son Isaac the rights to lumber and pasture in the area if he kept it weeded and "smooth from hogs and digging."

Child had arrived in 1698, fleeing the tyranny of an English lord, and been granted a large area including the bluff land known as Strawberry. Here in 1707 he began to operate a ferry and laid out the town of Childsbury, with property eventually set aside for a fortress, college, free school, church, and market square. Only the last three ever became a reality, and though the market and a related racetrack continued to operate for some time, the self-sufficiency of surrounding plantations made such towns needless, and Childsbury soon disappeared.

Built in 1725, this chapel was to serve those parishioners who could not make the distant trip to Biggin Church near Moncks Corner. As Rev. Dalcho pointed out in connection with Strawberry, a chapel was for convenience only and possessed only "the parochial rights of baptising and burying, but have neither Rectory nor Endowment." Services were held at the parish church only on major holidays. A hundred years after its completion, however, Strawberry had replaced the ill-fated Biggin Church and was enjoying the privileges thereof.

Not to be forgotten, of course, is the ordeal of Catherine Chicken. As a child of seven she was tied to a tombstone here by the schoolmaster who boarded her. Abandoned, she was rescued after a night of terror by a faithful servant. Her tormentor was drummed out of town, but one side of Catherine's face remained drawn and misshapen as a result of the experience.

Christ Episcopal Church

Built ca. 1726, Christ Church Parish, Charleston County, National Register.

In the 1950's historian Anne King Gregorie of nearby Oakland Plantation wrote of her Christ Church Parish with a thoroughness and ease that

The cupola reflects the tiny Christ Church's turbulent history. The original building had no such decoration but after being burned by the British one was added to the new roof. In 1800, with membership at its lowest and the church destitute, the cupola was removed when repairs were made. In 1835 treasurer Thomas Barksdale replaced it. In 1843 it was taken down and made higher. This cupola that we see today from the roof was apparently the only part of the building not destroyed by Union troops. The cross was added with the boxed eaves and new stucco in 1923.

could only incite unchristian envy from neighboring parishes. Designated simply as "Southeast of Wandoe river," the parish she describes so well was one of ten created by the Church Act of 1706. The following year a wooden building, 40' × 24', was built. When this building burned in 1725, plans already in progress to build the present church were quickly put into execution. At that time the population of the parish was listed as 470 freemen and more than 700 slaves, with parishioners described as "sober, industrious, and regular attendants on public worship." Despite this, only 30 communicants are recorded.

The new brick church was completed in 1726 by builder John Metheringham, who received £500 for his services. The interior was not completely finished at that time, and changes were made not long after, with the traditional south door becoming a window and the pews being altered greatly. For the most part the members of the small congregation were poor, and maintaining the services of a minister on a regular basis was almost impossible. For its first years Miss Gregorie records a sad recital of burnt schools, fever, and delinquent payments, including a 1738 report by "professed house carpenters" that the parsonage had rotted beyond repair.

These difficulties were minor, however, compared to what lay in store, for the church was burnt during the Revolution and had to be rebuilt within its original walls. To make matters worse, the need for a chapel in Mt. Pleasant led to heated debate and a lawsuit by Thomas Barksdale during the 1840's. Union troops gutted the building during the War Between the States, and once again it had to be essentially rebuilt. Services were discontinued in 1874, and the small church became, as Barksdale had predicted, the stepchild of its own chapel. With congregations following the shifts in population this was not an unusual occurrence, and the abandoned building suffered. During the first decade of the new century, the road came so close to the structure that the wagon hubs cut a groove in one corner and grazing cattle knocked many of the tombstones over. In 1923 the area was fenced and a renovation effort begun; in 1954 a determined congregation separated from the Mt. Pleasant chapel and full services began again at Christ Church.

Fenwick Hall

Built ca. 1730, Stono River, Johns Island, St. John's Parish Colleton, Charleston County, National Register.

In 1712, George I of the German house of Hanover came to the throne of England, and Paul de St. Julien named his plantation in honor of the monarch. But his gambrel roof for Hanover is not typical of what has come to be called Georgian architecture. In England the Georgian period accomplished a break with previous architecture by the almost undebated adoption of classical elements of design. It had been popularized by Sir Christopher Wren and Lord Burlington, who both drew heavily on the work of the sixteenth-century Italian architect, Andrea Palladio. By 1715, the use of builder's guides, handbooks, and English-trained craftsmen had spread the new architecture across the Atlantic, and it continued to dominate American architecture even after the Revolutionary War.

Built in 1730, the original section of Fenwick Hall is a fine example of the symmetry and grace of Georgian design; it reflects a desire for order and proportion represented in Mulberry and brought to full flower at Drayton Hall. We must assume that it was the intention of the designer, if not the conscious wish of John Fenwick, that his building at the water table measure exactly one and one half times wider than it is high. In addition, the distance from the center of the house at ground level to the corner of the cornice is the same distance as the height of the house to the top of the trim on the roof.

The shallow pitch of the hipped roof, the palladian windows, and the relatively unadorned entry were trademarks of what would have been considered a "middle class" dwelling in England. Here the solid brick building, with its finely paneled interior, was a "castle" for its owner. Descended from one of the "Red Sea Privateers," John Fenwick prospered and left the home to his son, Edward, who appeared in 1747 as a member of His Majesty's Council. Edward's first marriage to Martha Izard left him without an heir, but Thomas Drayton's daughter, Mary, provided him with sev-

eral. At the time of this second marriage in 1753, two flanking buildings similar to those built at Drayton Hall were added. The one still standing served as a coachhouse and the other as a stable for Edward's famed thoroughbreds.

An obituary from 1775 states that the Fenwicks had returned to England to see to the education of their sons, Edward and Thomas, who are both mentioned again in the Confiscation Act of 1782. Indicted for loyalist sympathies and involved in a family lawsuit, young Edward sold the house in 1787 to his cousin, John Gibbs. Gibbs added the north wing with octagonal ends, a form associated in England with the Adam brothers and popularized here by Charleston's Gabriel Manigault. We will see this influence later as an integral part of initial construction.

During the Revolution the house was occupied by both British and American armies but escaped relatively intact. By the 1930's, however, it was neglected to the point where sashes, shutters, and doors had to be replaced.

A classic example of Georgian design is this well-proportioned hall built by John Fenwick. The main building dates from 1730 and the octagonal-ended addition was built just after the Revolution. The wing to the far left is from our own century.

The much celebrated Drayton Hall is seen here from the land side. The massive Palladian portico is thought to be the first of its kind in America.

Drayton Hall

Built ca. 1738, Ashley River, St. Andrew's Parish, Charleston County, National Register, National Historic Landmark, Open to the Public.

Described by a neighbor as "Mr. Drayton's Palace" and by a modern enthusiast as "the finest early Georgian house in America," John Drayton's country seat has consistently inspired awe and praise since its completion in 1742. Though its architect is not known, the house is clearly the product of at least a master joiner working from the suggestions of a major design book. Several such books show similar Palladian elevations with distinctive two-story porticos, but the design for the "fox" overmantel is obviously taken from plate 64 of Kent's *Designs of Inigo Jones.*

The existence of the finely proportioned Fenwick Hall attests to the presence of such skilled artisans, but here the scale is much grander and the interior far more ambitious. Even where later Federal mantels have been substituted, great Georgian overmantels remain, and the cornices and doorways are all heavy and richly carved. Mahogany and yellow poplar ornamentations were left with a natural finish to contrast with walls originally painted in eggshell or marbleized. False doors were constructed for the sake of symmetry. Interior shutters, originally keyed into the sashes, controlled light. Ceilings were sculpted with great care, with one sculpted in wet plaster.

The floor plan is that of an English manor house, with the main entrance into a great hall or

Throughout Drayton Hall delicate carvings of mahogany were left with natural finish to contrast with the painted cypress paneling of the walls. Over the windows of this drawing room are mahogany garlands. Executed in wet plaster, the hand-molded ceiling is the only one of its kind in America.

Serving as both dining salon and ballroom, this second story hall was the center for entertaining at Drayton Hall. The great Georgian overmantel with broken pediment and fluted pilasters of the walls are found throughout the house. The door to the right opens onto the second story of the portico.

parlor. Two drawing rooms open off each side and a stair hall at the rear leads to an identical arrangement on the next floor. We can only speculate on the exact use of each room, but the upstairs hall was apparently the ballroom and center for social gatherings. We find a description of Drayton reclining in state while slaves fan with peacock plumes. It is of equal interest to imagine the activity that went on behind the scenes. A secret circular stairway winds through the center of the house, giving access to each floor and the basement, which along with the flanking kitchen and laundry, must have been the scene of great domestic activity.

Drayton Hall reached the twentieth century relatively unchanged. Neither plumbing nor electricity had been installed, and except for the window sashes destroyed by a hurricane, the Federal mantels, replacement of the roof, and the Victorian shingles in the portico gable, there has been no significant alteration from the original. The interior has been painted only twice so that the carving remains distinct.

The builder of the house, Royal Judge John Drayton, was considered at his death to be one of the wealthiest men in the colony, owning rice-growing properties and over 500 slaves. Drayton Hall was his country seat and not a working plantation in the strictest sense. From here he could manage his other estates, enjoy easy access to Charleston, and vie with his neighbors in a social circle already famed for its extravagance.

One Drayton memorialist left an unattractive description of Royal Judge John Drayton which does not appear to have been refuted: "Such was his character, he lived in riches—but without public esteem. He died in a tavern, but without public commiseration." His great fortune and the debts he had accrued were split between the descendants of four wives. He left his eldest son William, of Revolutionary War fame, practically disinherited. Nevertheless, the house remained in the family for the next two centuries.

Perhaps the most exceptional fact about Drayton Hall is that it has survived at all. Though it has almost become a cliché to list the litany of fire, storm, war, and earthquake that have taken their toll on these plantation houses, Drayton Hall is in fact the only one still standing on the Ashley River. Spared by the Union forces because it was in use as a smallpox hospital for Negroes, it was almost scrapped for its bricks during Reconstruction. Only the mining of phosphate gave the family the minimal wealth needed to preserve the status quo. But throughout it all "the venerable villa" has remained a mecca for visitors. As Ann Vanderhorst did after the War, we may look in vain for "the choice exotics" that once decorated the grounds, but this house continues to possess an austere majesty that seems anything but fragile. Drayton Hall is held by the National Trust for Historic Preservation and is open to the public.

Middleton Place

Developed ca. 1740, Ashley River, St. Andrew's Parish, Dorchester County, National Register, National Historic Landmark, Open to the Public.

Pleasing complements to Drayton Hall are nearby Middleton Place and Magnolia Gardens. These also are open to the public, and together, these three offer a composite picture of the grandeur that was commonplace to the locale.

The history of the dwelling at Middleton has

A portrait of Czar Nicholas I hanging in the Middleton music room reminds the visitor that the second Henry Middleton served a decade as minister to Russia. The Empire furnishings were made in Philadelphia (ca. 1820). The Aubusson carpet is mid-nineteenth century. The duet stand is English made (ca. 1815). Possibly painted by Greuze, the two-centuries-old "Girl with Fingers in Her Ears" still lightly exercises her role as resident music critic.

The garden of Middleton Place and the remaining South wing of the house are seen here with the Ashley River beyond. The main axis of the ground's design passed through the entrance gates, the central hall of the original house, and then between the Butterfly Lakes. The formal garden is to the left. The rice mill pond is to the right.

been the source of some controversy over the years. A sketch of the original that has recently surfaced seems to support the following thesis: As early as 1705 John Williams built a large three-story brick house. After his daughter's marriage to Henry Middleton, a pair of attractive flankers were added in about 1755. All were burned at the end of the War Between the States, and the earthquake of 1886 toppled the ruins except for the bachelor's quarters, which were extensively rebuilt at the beginning of the century in a romantic fashion unlike the original.

The curvilinear gables now in place resemble those of St. Stephen's and Pon Pon chapels, but the use of these features even in the eighteenth century constituted a revival of sorts. A Jacobean renovation optimistically planned for the house in the midst of the War would have been an even greater departure from the original Georgian structure, however.

No such misfortune has befallen the resilient garden, for it remains as it was originally laid out, a testament to the symmetry and order that we see expressed in the architecture of Drayton Hall. The inspiration at Middleton Place was the famed European gardens of the time, especially the work of the French landscaper Andre Le Notre. The main axis of the design runs from the entrance gate through what was once the central hall of the house and on between two butterfly lakes down to the river. The garden itself is at right angles to this line and is divided internally by precise paths and alleys that offer the visitor alternating moments of roomlike privacy and the sudden vistas of the river. Still unsurpassed today, this effort is all the more remarkable when we consider that it is the earliest known landscaped garden in America.

The builder, Henry Middleton, had acquired the property as part of his wife's dowry, but he became immensely wealthy in his own right. By the time of the Revolution he was probably Carolina's largest landowner, holding 50,000 acres and 800 slaves. His English grandfather, Edward, had set-

tled here after living in Barbados, and from a Goose Creek plantation began a family tradition of planting and of public service that continued uninterrupted for generations. His son Arthur was president of the convention that overthrew the Lords Proprietors. Arthur's son Henry, the builder, was President of the First Continental Congress, and his son Arthur signed the Declaration of Independence. In the next generation, Henry tended his beloved garden during the lull between wars, but his son Williams signed the Ordinance of Secession. After the War, Williams returned to a house in ruin and a garden run riot.

Magnolia Gardens

Developed ca. 1840, Ashley River, St. Andrew's Parish, Charleston County, National Register, Open to the Public.

The original seat of the Drayton family was not Drayton Hall, but the neighboring Magnolia Plantation. In the first decade of settlement, Barbadian Thomas Drayton, Jr. arrived and married another Barbadian, Ann Fox. Receiving the present site as a dowry, he built here one of the first mansions of the colony. According to a historian of a later generation:

> The house at Magnolia had a kitchen and office underneath and ample hall; and chambers both on the first floor and in the dormer story; and the ceilings of the first floor were remarkably high, being about 18 feet. The Hall was decorated with pilasters and other ornaments of stucco work.

Thomas' son John Drayton, who built Drayton Hall, did not inherit Magnolia, but before his death he owned both and left this property to his son Thomas. The first house was destroyed by fire and was replaced with another, which was subsequently burned by Union troops. The present structure was moved to the site in 1873 by Rev. John Grimke-Drayton.

The magnificent informal gardens at Magnolia are the legacy of Rev. Grimke-Drayton. As a young man he had been studying for the ministry when his brother was killed in a hunting accident, leaving him heir to a vast fortune that included Mag-

To placate a homesick wife, Rev. Grimke-Drayton worked to "create an earthly paradise in which my dear Julia may forever forget Philadelphia and her desire to return there."

nolia. He completed his studies, married Philadelphian Julia Ewings, and settled on the estate. While serving as an Anglican priest, he began to develop the gardens. In 1843 Rev. Grimke-Drayton imported numerous specimens of *Camellia Japonica,* and in 1848, *Azalea Indica.* After contracting tuberculosis in 1851, he retired for a period and devoted himself full-time to the gardens. He disregarded the formal French influence that dominated neighboring Middleton, and adopted a "Romantic" treatment. With an emphasis on the natural, paths wound beneath the magnolias through an apparently random planting of azaleas and camellias.

Rev. Grimke-Drayton was the nephew of the notorious Grimke sisters, abolitionists who had the distinction of starting riots both north and south of the Mason-Dixon Line. Early in his life he was strongly affected by their teaching. He referred to his 300 slaves as his "black roses," and in spite of legal prohibitions, he taught them to read and write. After the War, however, he found himself no better off than his less enlightened neighbors, and was forced to sell all of his properties except for 390 acres of Magnolia. In 1870, he opened the property to the public. Baedeker's travel guide listed it, along with the Grand Canyon and Niagara Falls, as one of the three greatest attractions in America. Still operated by the family and open to the public, the garden continues to receive worldwide acclaim.

Oakland

Built ca. 1740, Copahee Sound, Christ Church Parish, Charleston County, National Register.

Oakland was originally settled for John Perry of Antigua by his agent John Motte. Perry named his plantation after his native Irish parish of Youghal. At his death his daughter Mary came with her husband to America. Shortly afterwards, in 1740, she sold the plantation to Captain George Benison. The following year the Captain gave the 982-acre tract to his son George Benison, Jr., who left it in 1748 to his son William, with the provision that his other son, Richard, would inherit it if there were no other heirs. Both sons died soon after, and in 1755 the property passed from the provost

The unusual angled fireplace was originally built of tabby brick. Seen also in the design of Hanover are a window and door that open into a stairhall rather than the outdoors.

Neither the piazza across the far side nor this portico are original to this old-world design; an inspection of the attic suggests that the piazza came as a later concession to the climate. When the oak avenue was planted, this portico gave the house a new entry.

marshal to Thomas Barksdale. This chain of title suggests George Benison, Jr., as the most likely builder.

The small gabled-roof house is similar in some respects to Hanover. Here, however, the brick for foundation and chimney were made of tabby. The chimneys were built within the walls, rather than on the outside, and the fireplaces were angled in the front rooms. As with Hanover, the rear contains a stairwell separated from the remainder of the house by a second sidelighted entry.

After the War, the house, renamed Oakland, passed to Philip Porcher and from him to his daughter, Mrs. Ferdinand Gregorie. Historian Anne King Gregorie was reared there and returned there to stay during the Great Depression.

Though the owners never enjoyed the kind of prosperity that made it possible for similar houses to be replaced or expanded beyond recognition, the grounds, with dairy, tabby smokehouse, and original kitchen suggest a satisfying self-sufficiency. Inscriptions scratched into the windowpanes by both residents and visitors over the last two centuries imply domestic contentment, warm hospitality, and related sentiments: "March 10, 1773. Exalt Jehovah our God" wrote a John Wesley missionary. "If you my love do accept . . ." began a hopeful suitor.

Snee Farm

Built ca. 1754, Horlbeck Creek, Christ Church Parish, Charleston County, National Register, National Historic Landmark.

Appearing to be little more than a farmhouse, Snee Farm cottage is one of the most solidly built homes we will encounter. Great sills beneath the house rest on sturdy brick pillars, and all seem as sturdy as when they were built in 1754.

The builder, Charles Pinckney, was the son of an Indian trader and provincial office holder who died in relative poverty. Young Charles was supported in his studies by his uncle Charles Pinckney, husband of Eliza Lucas, and went on to prosper as both lawyer and planter. Active in the early days of the Revolution, he and his brother-in-law Miles Brewton were too conservative to win elections to

Washington had breakfast here at Snee Farm, Mr. Pinckney's country estate. Deceptively small and simple, the home of this constitutionalist shows solid craftsmanship—modest in its exterior, but heavily framed and surprisingly polished within.

This is the earliest entry hall we encounter, but in the years following entry halls became the norm. The paneling here runs in two directions: horizontally on the wainscot, and vertically above.

the First Continental Congress. When Charleston fell Pinckney swore his loyalty to the King before his death in 1782.

As was often the case, the builder's son, Charles, was a more committed republican and spent time aboard a British prison ship. His real contribution to the cause of liberty, however, came in the years that followed. Because of the Revolution he had not been educated abroad like his notable cousins, Thomas and Charles Cotesworth, but was tutored at home. Nonetheless, he made a deep impression on the political thought of the day. At the age of 27 he submitted his "Pinckney draught" to the Constitutional Convention, several fundamentals of which were adopted. He is considered by some people to be the most influential contributor to our country's Constitution.

Pinckney also served four terms as governor, and was a U.S. Senator, minister to Spain, and a member of the State Assembly. Initially, his strong support of Federal government won him the disregard of fellow aristocrats and the nickname "Blackguard Charlie." From the beginning, though, he was outspoken in defense of slavery, and in later years this issue would become more central to his thinking, and the powers of the national government less so.

Although his father's actions had caused Charles to lose a portion of his inheritance, he was still a wealthy young man and continued to prosper in the years following the Revolution. While he had a grander home in Charleston, he appears to have considered this simple house proper enough for greeting General Washington in 1791.

Motte House (Mt. Pleasant)

Built ca. 1755, Charleston Harbor, Christ Church Parish, Charleston County, National Register.

Like Charleston's Rhett house, Jacob Motte's plantation house is surrounded by other residences. Its name, Mt. Pleasant, has been adopted by the entire community, and the relative spaciousness of the old village remains.

The original building is thought to have consisted of a simple two rooms above, two rooms beneath and a hipped roof. There was a shed off the back and a piazza across the front and part of the southern facade. At the beginning of this century, R. H. Pinckney did major renovations in the "high Victorian" style of the day, adding a second story to the rear and placing a columned entrance on the street side. A new entry hall was created at this time and much of the interior was replaced, but some of the original woodwork remains: baseboard, upstairs mantels, and a strip of gougework

Shown here is Jacob Motte's Mt. Pleasant house seen from the harbor side.

Serving a prosperous parish, the Biggin Church survived two wars before finally falling victim to a forest fire. Only a portion of two walls stands to suggest the dimension and grace of the original.

that appears similar to the wainscoting of the Mottes' Charleston house.

The house's builder, Jacob, was treasurer of the colony for 27 years and owned several plantations. In *The Dwelling Houses of Charleston,* the Smiths quote his obituary: "His corpse was attended to the grave by a very considerable number of the inhabitants who were indeed real mourners." This fine testament they supplement with a mention of some of his and his wife Rebecca's 19 children and their prominent sons-in-law, Thomas Lynch, William Moultrie, and William Drayton.

Built high on a bluff facing the prevailing breezes of the harbor, the house with its 250 surrounding acres may have been considered a health retreat. Hoping to restore her ailing husband's health, Eliza Lucas brought Charles Pinckney to the house where he died in 1759. Within a few years the neighboring property was divided and promoted as being safe from disease. In 1808 Andrew Hibben did the same with Mt. Pleasant Plantation, and others followed suit.

The geography which prompted this subdivision also made the plantation ideal for military purposes. In the yard a faint impression remains of the breastworks from which the colonists faced the British before the battle of Ft. Moultrie. Later the British used the house for their headquarters. Cornwallis visited and Gen. William Moultrie reported here while on parole at nearby Snee Farm. A copy of Moultrie's forceful reply to British offers of a commission and land in Jamaica is displayed in the house. The note ends with the ironic remark, "Think better of me."

In this century Miss Petie McIver lived in the Motte house and left behind an entertaining and inclusive record of the Mt. Pleasant community. To one side in the yard stands the plantation's original kitchen which housed Scipio DeVeaux at the time of Miss McIver's purchase. He informed her that she may have bought the big house, but she hadn't bought his house. He lived on in the kitchen until his death many years later.

Biggin Church

Built ca. 1756, St. John's Parish, Berkeley County, National Register.

Called Biggin Church after nearby Biggin Creek, this is the parish church designated for St. John's, described by Dalcho as "a pleasant and healthful part of the country, where the Planters were generally, good, sober and teachable people." In 1711, Sir John Colleton donated the three-acre site, and Col. Thomas Broughton furnished the inside, "all of cedar, and at his own expense." The first building burned accidentally. The present one, now in ruins, was begun in 1756. As was sometimes the case, the strategic highground at the corner of three roads served an army as well as parishioners. The British stored powder there and, when they retreated, they burned the Church.

It was rebuilt soon after. In 1819, Dalcho wrote that the parish had two prospering glebes, and a report following the Civil War lists $10,000 in stocks and winter and summer parsonages for the ministers. Most of the church was lost in the War and the interior was vandalized and burned. The abandoned building burned again in 1890 when a forest fire swept the area, after which much of the walls were pilfered for brick.

Pompion Hill Chapel

Built ca. 1763, East Branch of Cooper River, St. Thomas and St. Dennis' Parish, National Register, National Historic Landmark.

As early as 1703 this community of mostly Huguenot planters had erected a small cypress church which in turn became a Chapel of Ease for St. Thomas and St. Dennis parish. In 1763 this building was replaced with the present structure.

Dalcho gives the dimension of the building as 48′ × 35′ and adds, "This was expected to cost at

Pronounced "Pumkin," the Pompion Hill Chapel has been called "a miniature Georgian masterpiece." The site is a bluff on the east branch of the Cooper River.

least £600." After final arrangements with the workmen, the total he gives is £3000 for the Chapel and £1000 for adorning the interior, figures which if correct suggest a gap between estimated cost and final payment.

The renowned brick maker Zachariah Villepontoux supplied the bricks from his Parnassus Plantation on Back River. He is usually credited with the design, which borrowed some details from St. Michael's in Charleston. Villepontoux carved his initials in the bricks by both the north and south doors, and William Axson—who was probably responsible for the brick work—signed his name along with a masonic emblem.

Exceedingly well-constructed both inside and out, the building appears today in an excellent state of preservation. This has not always been the case, however, for in 1842 Dr. Irving reported that only the sudden "hand of piety and affection" had saved the building from abandoned neglect. Similar hands have aided twice more during our own century.

Lewisfield

Built ca. 1774, Cooper River, St. John's Parish, Berkeley County, National Register.

Sedgewidge Lewis dowered his daughter with this property in 1767, and seven years later his son-in-law, Keating Simons, erected this fine example of traditional Low Country architecture. Simons, the grandson of the builder of Middleburg, chose for a layout two floors of four rooms each, divided by a central hallway. The dwelling was placed on a high brick foundation with a piazza shading the exposed front, and the upstairs bedrooms were no longer wedged beneath the eaves, but given more than full headroom.

Stoney, Simons, and Lapham point out that by the last quarter of the eighteenth century men were designing buildings for comfort during the hotter months. The country was still considered healthy during the summer, and it was not until about 1790 that the planters would begin to retreat from "the country fever" to the relative safety of the pineland, seashore, town, or distant spas. By then the type of building seen here with its boxlike utility and broad piazza had established itself as probably the most familiar plantation design. The stair tower disappeared, piazzas lengthened, and

The broad brick steps of Lewisfield flare outward and end in solid cylindrical newels. Judging by old photographs, this West Indian feature was common in St. John's Parish.

Framed by rich natural cypress, this doorway leads to the dining room of Lewisfield.

an occasional wing or odd room was added, but the basic floor plan remained the same. Personal expression was limited primarily to the decorative details of the interior.

The Revolution, which began shortly after the completion of Simons' house, found the builder quite literally in the thick of things. He was captured at the fall of Charleston and was paroled to Lewisfield, which the enemy was using as a landing. In 1781, Col. Wade Hampton, paid a spur-of-the-moment visit to the plantation and discovered a British raiding party. He took seventy-eight prisoners and burned two boats. Knowing he would be implicated, Simons went off to fight "with a rope around his neck." The penalty for breaking parole was hanging, a fate he avoided successfully, fighting with Francis Marion and rising eventually to the rank of Brigadier Major. His warm relationship with Marion continued after the war, and the plantation of the heirless Marions was bequeathed to Simons' son.

The fine woodwork of this Lewisfield room includes an original built-in corner cupboard. The double row of dentil molding in cornice and mantel is distinctive.

Heavy untapered columns of a West Indian piazza encircle three sides of this small brick house.

Otranto

Built ca. 1790, Goose Creek, St. James Goose Creek Parish, Berkeley County, National Register.

Judge H. A. M. Smith dates the construction of Otranto on the far side of the Revolution, with Dr. Alexander Garden as builder, while Stoney, Simons, and Lapham place it on the near, giving the honor to the doctor's son, Major Garden. Though Linnaeus gave the doctor world-wide recognition with the naming of the Gardenia, it would be nice for sentimental reasons alone to credit the house to him as a more local and concrete reminder of his accomplishments.

A far cry from the "ghostly castle" of Walpole's novel, from which the house takes its name, this Otranto is a small piazza-bound dwelling reminiscent of the West Indies. The great untapered columns and scored stucco exterior give a sense of solid permanence that is accentuated by the site. Originally called Yesho, the property was settled by Arthur Middleton in 1768, and judging by the grand oaks and the view, the high bluff was possibly the site of an earlier house. It is known that Dr. Garden cultivated an extensive and elaborate garden beneath the magnolias and built a miniature temple to house his library.

One of the colony's leading naturalists, the doctor carried on an extensive correspondence with the great taxonomist Linnaeus, and was also in contact with England's Royal Society. For the Society he conducted numerous experiments attempting to vary the crops produced in Carolina, in the hope that diversification would guarantee a firmer foundation of wealth for both colony and Crown. When the Revolution began, Dr. Garden threw his lot in with the latter and returned to England.

Dr. Garden's son remained, fought under Light Horse Harry Lee, and wrote *Anecdotes of the Revolution.* By one account Major Garden was kidnapped from his home by the rebels whom he chose to join. In any event, relations with his Loyalist father remained strained.

Before 1800, the house served as a parsonage for the Goose Creek Church, and then, for a brief period, was rechristened Goslington by Philip Porcher. Famed for its hunting, Otranto was operated as a hunt club almost to the present, and while serving this function was gutted by fire, but quickly restored. It can be easily viewed today, for the community of Goose Creek has expanded quite suddenly, and the house is but one of many on the winding streets of Otranto Subdivision.

Halidon Hill (Quinby House)

Built ca. 1792, East Branch of Cooper River, St. John's Parish, Berkeley County, National Register.

In 1681 John Ashby received a grant to "the southernmost side of the eastern branch of the Cooper River" and was given the title of "Cacique" with the entitled baronies to be given "as he required." He appears to have made arrangements during the following year for construction of his plantation mansion house, sending over from England Henry Russell with funds for his "Lodging . . . cloathes and all manner of carpenter's tools fitting for his works." Ashby's son John came over to manage affairs and inherited the plantation that is named after the family's English estate of Quenby Hall. (The name was corrupted to Queenbie, Queenbee, and the Quinby used today.)

Ashby married the sister of Lt. Governor Thomas Broughton, and the plantation passed by marriage to the Shubrick family. The buildings were the site of a bitterly fought battle between Gen. Coates and Gen. Sumter. The house was destroyed during the Revolutionary War or soon after. In 1792, the property appeared in an advertisement as the following: "Quinby Plantation for sale, about 26 miles from Charleston containing 158 acres of rice land in the very best pitch of tide.

Dancing nymphs of this finely preserved Adam mantel are typical of Halidon Hill's Federal embellishments. The bell rope to the right still works. The English secretary contains a variety of fine porcelains.

1,045 acres of high land contigious to several fine springs of water."

Roger Pinckney, the son of the colony's provost marshal bought it that year and apparently built the house soon after.

The house is an unusual variation of a familiar floorplan. The second story, though reached through a stair tower, is only one room wide, and large closets with windows are let into the space beneath the shed roofs. More closets, an unusually wide entrance, a graceful mahogany banister rising to the third floor, and the luxury of double bell ropes ringing in three locations indicate that an unusual degree of personal thought and care went into the house's design. The interior is given a full complement of Federal embellishments. Dentil cornice, wainscoting with terra cotta panels and marbleized baseboards, brass rosettes, and the dancing nymphs of an Adam mantel all suggest a

Beginning two flights below, this mahogany banister makes a curiously graceful bend here at the third floor.

This spacious dining room at Halidon Hill is handsomely complemented by early furnishings in the Chippendale and Hepplewhite styles. Two Rockingham tea sets are seen as well as Spode-Copeland china. The curtain rods are thought to be original and when the house was moved in 1954 made the long journey from Quinby in place.

post-Revolutionary return to full planting productivity and prosperity.

Isaac Ball bought Quinby in 1816, and it remained in his family through the War Between the States. It is said that Union soldiers were told by the freed slaves that the missus had been good to them. And so instead of burning the house, the soldiers danced before the portrait of Jane Ball.

In 1954 the Quinby house was moved four miles through the woods and down a narrow highway to its present site at Halidon Hill. This latter property was originally associated with Middleburg and was willed by the third Benjamin Simons to his daughter Catherine who married William Hart. Eventually it came into the possession of John Coming Ball, who reunited the tract with Middleburg for a while.

Celebrated in a poem by Sir Walter Scott, the original Halidon Hill was a fourteenth century Scottish battle site. Writing in 1842, Dr. Irving claims that owner William Ball chose this name for his recently purchased property. "A Hill" only by Low Country standards, the gentle rise covered by a fine expanse of lawn makes a beautiful riverside home for the orphaned Quinby house.

Hyde Park

Built ca. 1799, East Branch of Cooper River, St. John's Parish, Berkeley County.

Built by John Ball in 1799 for his wife Jane as "a nice place to have quilting parties and go marooning in the spring," this small house offered a prospect of the Cooper River denied to the landlocked main house of Kensington. Since the cottage was intended only as a retreat, a single economical chimney was placed at its center. The chimney begins in the basement with kitchen, fireplace, and oven, and on the main floor two small cater-cornered fireplaces back up against the large one of the main room. The interior is finished simply with high plaster walls broken only by a wide, beaded baseboard and a chair mold, but its beaded corner post, H and L hinges, and double entry doors suggest an earlier age of building.

Added perhaps at a slightly later date, the back sheds and small stair tower are reminiscent of nearby Quinby, now Halidon Hill, a resemblance that must have been particularly striking when its piazzas extended around all three sides.

John Ball, the builder, was the fourth in line of a vital and colorful family that had settled early in the Strawberry area of the Cooper. Although only 16 years old, he had inherited the responsibilities of his father Elias, Sr. "Manly and self-reliant," he had managed well. Three years later he was riding in Horry's Cavalry. In 1780, he married Jane Ball, and while living at Kensington, they had five children before she died of the "long decline." Remarried to Martha Swinton, he fathered 11 more children. Practical, with a good business sense and an eye for detail, he prospered. By 1817, when he died of bilious colic, he had acquired half a dozen other plantations, including Marshlands, and a fine and fashionable townhouse. "Charity and benevolence combined with prudence," it is written, "marked his character."

His wife Jane appears to have shared both her husband's prudence and benevolence. A record remains of the sale of poultry she raised for what was considered pin money, but she also found time to carry on an extensive correspondence with two wide-ranging sons. The lives of the Ball family are all well-documented by letters and diaries. A finely condensed version is presented in Ann Simons Deas's *Recollections of the Ball Family of South Carolina.*

Writing from Kensington, Jane Ball reported in 1798: "The house on the hill at Hyde Park is not yet finished, it is a pretty situation and will be a most delightful spring and summer residence, if the house was larger I should like much to stay there."

"The world forgetting and world forgot seems the motto for the Island," wrote a visitor to Kiawah in 1900. "Desolation and decay are everywhere, and yet it is beautiful."

Kiawah (Vanderhorst House)

Built ca. 1807, Kiawah River, St. John's Parish Colleton, Charleston County, National Register.

"The sound of the breakers is ever with us" wrote a visitor in 1900, describing the land abandoned to decay and desolation, a forsaken avenue with "beauty on either side, here with tangled masses of vines and foliage running riot." Despite the recent extensive development of Kiawah Island, remarkably little has changed to alter this picture. The oaks of the avenue have died, but their trunks remain to outline a vague indentation through the undergrowth. Of the piazza from which the observation was made, only the arcaded foundation remains, but the house with new roof and boarded windows awaits a planned restoration.

It was the Kiawah tribe that "once peopled these shadows" and recommended the site on the Ashley river for the Charles Town settlement. In 1699, the island was surveyed for privateer Capt. George Rayner. By 1739, ownership had passed to John Stanyarne, who had cleared much of it and was successfully cultivating indigo. He is often credited with building the house, but details of its design and the claim that the British had destroyed the island dwelling suggest a later date of 1807, the year that litigation ended and gave Stanyarne's son-in-law Arnoldus Vanderhorst clear title.

Vanderhorst had a long career as a public servant. Among other things he helped to establish the College of Charleston. A leading Federalist, he served as intendant (mayor) of Charleston and was eventually governor of the state. He had expanded the planting on the island, replacing indigo with cotton. His son Elias continued a diversified management, producing live oaks for ship building, palmettos for fortifications and shell for lime.

The house, probably built by the father or son, rises an austere two-and-a-half stories on a high brick foundation. It is hard to imagine it as Mrs. Vanderhorst's "white curlew" of earlier days, for no evidence of paint remains. Inside, a conventional central hall divides the rooms, which have alternating panels of dark and light wood and mantels fluted with inlaid urns.

Marshlands

Built ca. 1810, Cooper River, St. James' Parish Goose Creek (now moved to James Island), Charleston County, National Register.

Built by John Ball, Jr., in 1810, Marshlands has remained basically unchanged in everything but location. In 1961, it was loaded intact aboard a barge and floated from its site within the bounds of the Naval Base to a new foundation at Ft. Johnson on James Island.

This fine Federal house has the typical floorplan for its day, four rooms below with two more above; but the interior is exceptional. Each room is festooned with a different combination of gougework and the plaster ornamentation is exceptional. Of particular interest are the downstairs mantels: one depicts a river goddess holding sheaves of rice and cherubs leaning on farm tools, and another is said to duplicate a carving from a Roman tomb that was excavated in Turkey around the time that the house was built. In addition, combinations of rosettes, acanthus leaves, and stars line the cornices in an intricate display of molded putty adornments in the Adam style. No room is neglected, and even the porch gets a treatment that would have served well in many of the drawing rooms of the period.

Stoney, Simons, and Lapham suggest that the extensive use of gougework throughout the house may have been necessitated by the embargo prior to and during the War of 1812. The house is certainly pivotal in this sense, for in the years after its construction, the complex carving of local craftsmen would replace imported designs of sculpted putty. We will see in later houses such as Summit and Lawson's Pond grand explosions of sunburst ornamentations.

A similar house was built by John Ball, Sr., in Charleston, and the work for it was probably done by the same craftsmen. The two houses certainly mark a transition of this Ball family from the simple austerity of the earlier Kensington.

The builder, John Ball, Jr., was an eldest child. Born at Pompion Hill in 1782, he had been well educated and had considered entering the ministry, but passed up this vocation on his uncle's advice. "Marry Betsy Bryan and I will settle you at Commingtee." John and Betsy had five children before her death in 1812. Remarried to the widow of Thomas Simons, John had three more children. "Upright, firm, and just, but also kind and generous," his memorial reads. He died of country fever in June 1834.

Floated down the Cooper River and across the harbor to this new site on James Island, Marshlands now serves as offices for the South Carolina Wildlife and Marine Resources Department.

This small Church of St. Thomas and St. Dennis offers a "Greek Revival" contrast to the lower jerkin-headed buildings of pre-Revolutionary Anglican construction.

St. Thomas and St. Dennis Church

Built ca. 1819, St. Thomas and St. Dennis' Parish, Berkeley County, National Register.

Since the original 1706 boundaries of St. Thomas Parish encompassed the French settlement of Orange Quarter on the Wando, it was divided, and a second parish called St. Dennis was created for the Huguenots. At their request an Episcopal clergyman was sent to read sermons in French. This arrangement lasted for another half century, until the St. Dennis Chapel was replaced with one at Pompion Hill.

Completed in 1708, the original church of the parish of St. Thomas was at Cainhoy. This larger building burned "in a conflagration of the woods" in 1815 and was replaced four years later with the present structure. It is the only "new" church remaining from this post-Revolutionary period, and the high walls and full gable treatment are an interesting contrast to earlier designs. It has the dubious honor of being the site of the "Cainhoy Massacre," one of the more violent incidents of Reconstruction politics. Bullet holes were said to scar the ancient vestry building nearby.

Dean Hall

Built ca. 1827, Cooper River, St. John's Parish, Berkeley County, moved in 1971 to Huspa Creek in Beaufort County.

Built on a peninsula at the T of the Cooper River, Dean Hall and its numerous outbuildings were said to resemble a village more than a plantation, a logical description considering how self-sufficient most of these holdings were. Dean Hall had been settled by a family of Scottish baronets, the Nesbetts of Dean, the first of whom, Alexander, bought the property in 1725. By the time of the Revolution it had passed to his grandson, Sir John, described as exceedingly handsome, but having a "Haughty Stubborness of Temper." After a brief career in the military at Gibralter, he journeyed to Carolina to take over management of his inheritance. We know little of his life here, but he was mentioned in 1796 in *A History of the Turf* and was celebrated for his race against John Randolph of Roanoke, who defeated him in an exciting contest. The women's admiration, it seems, went to the "very elegant" John who "won the prize from beauty's eyes." Soon after the race he was married to Maria, the daughter of Waccamaw's great "worthy

Seven brick arches on each side support the columns of an encircling piazza at Dean Hall. It is said that the housebuilder, William A. Carson, could watch both his cotton and rice crops and the nearby slave quarters from the vantage point of his porch. The house now overlooks Beaufort's Huspah Creek.

New owners are furnishing Dean Hall once more in the grand style of planter-builder William A. Carson.

of the turf" Col. William Alston, and the couple then returned to Britain to stay. In 1821, he sold his 3,100-acre property to William A. Carson for $40,000, and the Yale-educated Carson built the house in 1827. Married to Caroline Petigru, daughter of the brilliant Charleston lawyer, James L. Petigru, Carson was quite successful as a planter, and his home was said to contain many fine works of art and rare books. The brick house, with its raised-cottage elevation and its living areas beneath an arcaded basement, was unusual for its day.

Carson's son, writing after the Civil War, reminds us that not everyone shared a grand vision of the planter's life. "My father, William A. Carson, was a rice planter who wore out his life watching a salty river, and died at the age of 56, when I was ten years old."

Young Carson sold the property in 1909 to Benjamin Kittredge, who turned its reserve into the world-famous Cypress Gardens. The house site was eventually bought for use as an industrial park, and the house was taken apart and rebuilt at Garden's Corner, near Beaufort. After years of labor, controversy, and disappointment, the rebuilt Dean Hall is almost completed.

The Blessing (Cedar Hill)

Built ca. 1834, Lynch's Creek and East Branch of Cooper River, St. Thomas and St. Dennis' Parish, Berkeley County.

In 1682, Thomas Ashe described Carolina as famous "for Salubrity of Air, Fertility of Soyl, for Luxuriant and Indulgent Blessings of nature." And of the many bold colonists who had hazarded the adventure of settlement, he singled out Jonah Lynch for having planted barley:

> Mr. Linch an ingenious Planter, having whilst we were there very good growing in his Plantation, of which he intended to make Malt for brewing of English Beer and Ale, having all Utensils and Convenience for it.

The Blessing was one of Lynch's two plantations, and is recorded as 780 acres "on the southside of the Eastern branch of Cooper River at a place called Wattesaw also, the Blessing . . ." Though it may have been, as Ashe wrote, a blessing of nature, the name is usually identified with the first of the ships to have brought the colonists. Jonah apparently married Governor Johnson's sister, and their son, Johnson Lynch, inherited the place in 1712. In that year it appeared as a ferry landing.

From there it passed to the Bonneau family and is mentioned by Mrs. Leiding as connected with the Laurens family.

Hunting at the Blessing before the War, William T. Sherman suffered an accident and spent several days at the house. This association later resulted in the family's Columbia home being spared from destruction.

The present house was built in 1834 by James Poyas and wife Charlotte Bentham. The son of a Beaufort merchant, Poyas appears to have held the property for only a short time, for it is advertised in 1837.

> At private sale vaulable Tide swamp plantation on eastern branch of Cooper River called Cedar Hill, part of that plantation formerly known as the Blessing, contains 150 acres under bank—rice land as fine as any on the River—610 acres High Land—2 good barns—a winnowing house, a large covered building put up for a Picker Machine also a dwelling house.

Built on a high brick foundation, the dwelling house is a functional country home with piazzas running down two sides. There is little embellishment on the interior or on the outside, which has only simple Doric columns and a rectangular transom above the door.

South Mulberry's builder, Dr. Sanford Barker, was an avid naturalist, but unfortunately kept no records of his work. Mrs. Leiding reports, however, that fellow scientists came for extended "botanizing" visits.

South Mulberry

Built ca. 1835, Cooper River, St. John's Parish, Berkeley County.

South Mulberry's existence as a separate plantation began in 1809 when Thomas Broughton willed the southern half of Mulberry to his son Philip S. Broughton. A survey of the property at that time shows no house, and it is assumed that the present structure was built in about 1835, by Dr. Sandford W. Barker who had married Philip's daughter Christina.

The Barker's "Home Place" has the double entry doors that we will see again, most especially in the Eutawville area. A large piazza extended around three sides of the house. Most distinctive is the basement foundation which is thickly built with windows guarded by diamond patterned strips. Remnants of Dr. Barker's garden remain.

South Mulberry was used for duck shooting early in this century. Large additions and extensive remodeling took place while the house was serving as a hunting lodge.

Boone Hall Slave Street

Built ca. 1843, Wapeckercoon (Horlbeck) Creek, Christ Church Parish, Charleston County, National Register, Open to the Public.

Before 1700, Major John Boone purchased his plantation on what was then called Wapeckercoon Creek. The Major was active in the politics of the day, and his family would continue to play an important role in the government of the colony and state. One daughter, Sara, was the grandmother of Edward and John Rutledge.

In 1817, the Boones sold the property to John and Henry Horlbeck, sons of Charleston's successful master builder John Horlbeck. The family operated a large brick and tile works on the grounds, as well as an extensive cotton plantation.

The slave cabins are thought to have been built in 1843 by four sons of the third generation. The presence of the brick works probably accounts for the large number of miscast and broken bricks that went into the irregularly bonded cabins. Reserved for the plantation's house servants, the tile roofs of these dwellings were a luxury, but the dirt floors, with wide brick hearths and sashless windows were the norm for slave quarters.

Also unusual, possibly even unique, are the elaborate diapers that decorate the brickwork in three of the cabins, and the round smokehouse beyond. Blue glazed brick were laid in single and interlocking diamond patterns, and seem to reflect not only the owner's wishes, but the pride of the craftsman as well.

A Horlbeck descendant left this description of the plantation beyond the big house:

> Miles of pasture upon which fine stock is raised . . . the gin houses, stables, barns and dozens of little cottages where the several hundred slaves have their home—not in a negro quarter but dotted about over the country, each with its little patch of land for the tenant.

The oak avenue and cabins of the Boone Hall Slave Quarters date from 1843. The intervening rows of pecans were planted early in this century when Boone Hall was the largest pecan grove in the world.

Gippy

Built ca. 1852, Cooper River, St. John's Parish, Berkeley County.

Built in 1852 to replace a burned home, Gippy is an example of the Greek Revival architecture that is usually associated with the Southern plantation. The columns were originally square, a common feature of "country" construction, but the newer round pillars pleasingly complement the front. The treatment of the gables and the even arrangement of windows gives the house a fine balance, and the front door, with the flat lintel and pilasters common to the style, is framed by unusual circular sidelights that draw attention to the entrance.

The property was part of Fairlawn Barony, granted to Peter Colleton in 1685, and it remained in that family until John White purchased it in 1818. This tract, "Gippy Swamp," was said to take its name from an old runaway slave who hid there in a hollow tree. White built a home which burned, and his son John replaced it with the present house.

The builder died of typhoid fever at the beginning of the War, leaving no man in the house but his 11-year-old son, J. St. Clair White. In the spring of 1865, he, his mother, and another companion faced a Yankee invasion alone:

> A few days later our scouts came through to warn us that the Northern Army was near, and before they could get out of the fields at the North of the Plantation the blue coats were literally pouring over the fence at the south and west. More like a mob, it seemed to me than an army . . . In a few minutes there was a mad scramble to capture or kill all the fowls and turkeys that were in sight. Mules were hitched to our light carriage which was loaded with bacon, and all movable supplies were taken. A few soldiers came into the house and carried off valuables, my father's watch along with the rest.

Despite this ominous beginning, the family's retainers came to their aid, and though there were "some trying times, nothing serious happened."

Samuel P. Stoney bought the house in 1895 and sold it to Nicholas Roosevelt in 1927. It was Roosevelt who restored the dwelling, substituting round columns for square ones, and introduced the herd of Guernsey cattle whose "Gippy" milk is still available today. As with Otranto, suburban sprawl has overtaken the plantation, and today the house is just one of many on the Gippy Swamp tract.

When the original home burned, neighbors lent their slave artisans and the Greek Revival house at Gippy was built "in record time." The gable end of the portico and house appear identical in this well-balanced construction.

In 1926 the addition of this portico gave the McLeod house a new entrance.

McLeod

Built ca. 1858, Wappoo Cut, St. Andrew's Parish, Charleston County, National Register.

Captain David Davis received a grant for this tract in 1703 and passed it to Samuel Peronneau in 1741. Peronneau's will provided that the plantation "be worked to support family until youngest child reaches twenty-one years of age or marries, then plantation to be sold and money divided equally." William Parker bought it through the master of equity and sold it to William McLeod in 1851.

The present house was built in 1858 as a conventional James Island cotton planter's home. During the War, the building was used as the headquarters of several Confederate units, and a sign on one of the upstairs rooms reads "Adjutant's room 2d."

Original to the plantation are five slave cabins and a garden at the house's former front. In 1926, a new portico was placed at the rear, for the advances of progress had rearranged the house's orientation. The creek is a part of the Intracoastal Waterway today, and the oak-lined drive that once led to the house is cut by a public road. Enough of the vegetation remains, however, to still shield the house and grounds from the buffets of the twentieth century.

Several of the early occupants at Fairfield, the oldest home on the Santee River, built additions onto the house. Amateur architect Gov. Thomas Pinckney lived here for much of his adult life, but apparently resisted the temptation to alter the house.

Santee River System

FORCED out of France by the revocation of the Edict of Nantes, French Huguenots settled in the Jamestown area of the Santee in 1689. "Since leaving France," wrote Judith Manigault, "we have suffered every kind of affliction—disease—famine—pestilence—poverty—hard labor." Less than a decade later John Lawson reported "seventy families seated on this River, who live as decently and happily as any Planters in these Southward parts of America."

English settlers joined the community. The marshes of the vast delta were diked and the planting of rice and indigo brought great wealth to the society of "French Santee."

In 1843 an anonymous Yankee traveler wrote:

> For many miles up and down the North and South Santee rivers, which are here separated by a single mile, are cultivated those deep, rich bottoms, annually flowered and inexhaustible in resource, which are the glory of the State. The lordly owners of these manors pass the winter months in superintending the affairs of the homesteads, gathering about them all those luxuries which minister to ease and pleasure, of which none better understand the value, or select with more taste, than do these descendants of King Charles' cavaliers, and entering with a zeal and alacrity into those rural sports which are the zest and glory of a southern country life. Finer horsemen, more skilled marksmen, on the plain or in the forest, hardier frames for pugilistic feats, or a quicker eye and prompter hand for a game at fence, the work cannot produce. They are generally men also of liberal learning and generous dispositions; frank, hospitable, and courteous; and, bating a tithe of that hot-blood chivalry upon which they are too apt to pride themselves, noble and humane in all their impulses.

Fairfield Plantation

Built ca. 1730, Santee River, St. James Santee Parish, Charleston County, National Register.

In 1822, Robert Mills visited Fairfield and gave this description.

> . . . The situation of Col. Pinckney's house commands an extensive prospect of what may be called the gold mines of the state, "namely rice fields. The Santee river glides majestically by silent and slow. A number of handsome country seats stretch along its banks, each fronted by rice fields. The prospect here is said to be enchanting in the spring when the young rice is shooting luxurious from its bed. Even at this time the verdure is beautiful in weeping willow, wild orange, sweet orange and mock orange, besides innumerable evergreens. Fairfield is handsomely laid out, possessing all that enchantment which a climate like this is capable of affording . . .

Almost a century had passed since Thomas Lynch had come up the Santee to the first inland bluff. On a high brick basement he built a house consisting of four rooms on the first floor and two more above on the river side. In 1766, Jacob and Rebecca Brewton Motte added two more rooms and rearranged the interior. Their daughter married Thomas Pinckney, and the house passed to his son, Thomas, Jr., who was Mills' host. To Fairfield this Thomas added the small porches and two wings. In such a manner, the building took on the symmetrical shape of the large wooden Georgian plantation house that we will see repeated so often.

As for the occupants, they all played prominent parts in the settlement of the country, but Thomas Pinckney bears special mention. The son of Eliza Lucas Pinckney, he was educated at Oxford and returned home after a 16-year absence to play a minor part in the Revolution: he was wounded and later captured. "Though not yet complete in anything," he was said to "show great promise." He went on to become Governor of South Carolina, our first Ambassador to the Court of St. James, Emissary to Spain, Vice Presidential nominee of the Federalist Party, and commander of U.S. forces from Virginia to Mississippi during the War in 1812. All his life he maintained an avid and scientific interest in planting and at least an amateur standing as an architect. He is thought to have designed several stately houses in Charleston and is known to have planned and built for himself a larger, finer, and less traditional house at neighboring El Dorado.

Hampton

Built ca. 1735, Santee River, St. James Santee Parish, Charleston County, National Register, National Historic Landmark, Open to the Public.

Hampton Plantation has had the good fortune to experience two sympathetic restorations during this century. The first, which could be called a loving and romantic interpretation, was carried out in 1937 by the poet Archibald Rutledge and is well documented in his *Home by the River.* The house had been in his family for 200 years. He recalls:

> When I came to where the gate used to be, I could hardly see the house for the tall weeds and the taller bushes. It was as if the blessing of fecundity had been laid on everything natural, and on everything human, the curse of decay.

Wearing sashes painted with the President's likeness, Harriott Horry and her mother Eliza Lucas Pinckney stood on the Hampton portico to greet Washington. Because her late husband had used the front lawn for a racetrack, the oak we see had apparently been planted with some misgivings. The President recommended that it stay, and it remains today the Washington Oak.

This is one of the few ballrooms still in existence outside of Charleston. Floor boards run unbroken for over 30 feet. The ceiling has been returned to its original sky blue.

Not to be discouraged, he set to work with the help of loyal plantation blacks, to restore the house and grounds, and in the process wove a fabric of fact and legend that still endures.

The second and more scientific restoration was recently undertaken by South Carolina Department of Parks, Recreation, and Tourism. After the plaster was removed, the interior of Hampton house was carefully rebuilt with some parts left uncovered so that visitors could see how the house was constructed.

In 1695, Huguenot Elias Horry arrived in French Santee, probably acquiring the Hampton land between 1700 and 1730. The house may have existed in 1736 when Horry died of "country fever," but family tradition claims his son Daniel built the house in 1750. (Containing remarkably similar mantels, paneling, and stairs, the Charleston house of Colonel Othniel Beale was built shortly after the fire of 1740, a date closer to the 1735 suggested by Stoney, Simons, and Lapham.)

The Hampton house began much as did Fairfield, with two rooms above four and a shed roof sloping toward the river. Sometime before the Revolution, a major enlargement took place. Two rooms were added above, a ballroom was built on one side, a corresponding two-story addition was built on the other, and a new roof covered the

This cross section of colonial construction reveals the mortise and tenon joining of beams that had been shaped with broad axe and pit saw. The ceiling height was apparently increased after construction began. Above we glimpse the false window in the attic space that was created.

whole building, creating a massive image. By 1791, the further addition of the hallmark portico increased the house's distinctiveness.

The portico, incidentally, has an architectural history of its own, for it is possibly the first appearance of the Adam brothers' design in the southeast. Daniel Horry, Jr.'s wife Harriott was the daughter of the renowned Eliza Lucas Pinckney. She had often admired the English actor David Garrick's performances and must have admired his home on the Thames called Hampton. Eliza apparently carried both name and portico design back to the new world. She made her final home with her daughter at Hampton, and together they greeted General Washington when he visited on his tour in 1791.

Though the plantation is linked to the Rutledges, Horrys, and Pinckneys, it enjoyed its greatest prosperity during the widow Harriott's occupancy, and she continued to manage the property until her death in 1830. The following observations on Santee society, recorded by a visitor to Hampton in 1804, offer an interesting counterpoint to the state of decay which greeted the returning Archibald Rutledge:

> Within their houses you meet great hospitality, the polish of society, and every charm of social life; and abundance of food, convenience, and luxury. It is impossible but that human virtue in such a situation, doing justice to those under him must feel himself lord of the earth.

Hopsewee

Built ca. 1739, Santee River, Prince George Winyah Parish, Georgetown County, National Register, National Historic Landmark, Open to the Public.

"We were all vastly pleased with Mr. Lynch," John Adams commented in 1774. "He is a solid, firm, judicious man." Fellow patriot Silas Deane was similarly impressed: "Wears the manufacture of this country, is plain, sensible, above ceremony, and carries with him more force in his very appearance than most powdered folks in their conversation. He wears his hair strait, his clothes in the plainest order, and is highly esteemed."

Sometime between 1739 and 1740, Thomas Lynch had chosen a high bluff on the North Santee River almost directly opposite his Fairfield home, and there he built a house thoroughly in keeping with his character. The original was apparently two stories of four rooms each on a scored, stuccoed brick foundation, with hipped roof and small portico facing the river.

Thomas Lynch, Jr., was born there in 1749. At the age of 15 he was enrolled at Eton, afterward attended Cambridge, and then studied law at the Middle Temple, London. In 1772, he returned home and was soon caught up in the events of the Revolution. He attended the First and Second Provincial Congresses and served as a captain in the First South Carolina Regiment. In 1776, he and his father attended the Second Continental Congress. Both men were ill at the time, and only the younger signed the Declaration of Independence. Lynch Sr. died on the way home. In 1779, Thomas Jr., sailing for France because of his worsening health, was lost at sea. As with others in the planter

Birthplace of Thomas Lynch, Jr., a signer of the Declaration of Independence, the Hopsewee house commands a fine view of the Santee Delta.

An unusual candlelight molding trims cornice and mantels at Hopsewee. A later addition, the wall at the right created a central entry hall.

clan, neither father nor son stood to gain economically by the Revolution, and both were suspicious of a "Republican" form of government. The younger Lynch saw it immediately as a threat to slavery, an objection that Charles Pinckney would address in the following decade. Nevertheless, they not only joined, but helped lead the struggle for independence.

Robert Hume bought Hopsewee in 1762, and after the war it passed to his son John who is supposed to have turned down a Scottish earldom in order to remain "Earl of Marshmud" in South Carolina. Hume's grandson, John Hume Lucas, inherited the neglected house and in 1846 extensively remodeled it. Also the grandson of inventor Jonathan Lucas, the new owner's addition of a double piazza across the front was a definite triumph of functionalism. The prevailing breeze crossing the wide expanse of delta made for a very livable dwelling.

The house is generously open to the public for a nominal fee.

Peachtree Ruins

Built ca. 1762, Santee River, St. James Santee Parish, Charleston County.

Thomas, the son of the Blessing's Jonah Lynch, had extensive rice plantings on both the North and South Santee rivers. A 1738 inventory taken at his death listed the following plantations under cultivations: New Ground, Brick House, Indian Field, Pleasant Meadows, Hopsewee and Peachtree.

Judging by the list of slaves, livestock, and utensils, most of these plantations were self-sufficient tracts (complete except for a main house). The inventory for Peachtree includes 37 hoes of varying types necessary for rice production, and enough coopering tools to suggest that barrelmaking was a major activity on the plantation.

The Thomas Lynch of the following generation had moved from Fairfield to Hopsewee, but since that property was sold in 1762, we can assume that by then he had crossed back over the Santee and completed his new home at Peachtree. The house burned in 1846 but from the ruins it is obvious that the house was substantial. Measuring 48′ × 55′, it was a full two stories high above a nine-foot basement. The stuccoed and scored brick walls are almost three feet thick at ground level, and unusually large openings indicate that the windows were nearly six feet high. Both entrances to the house received elaborate treatment, the water side having a set of rectangular stone steps, and the land entry semicircular ones with a landing paved with heavy red tile. It has been suggested that the house was laid out with the river entry opening into the larger of two front rooms and a large stairhall at the recessed rear entry.

A visitor of the day left this description:

> In its architectural style and general appearance—the massy materials of which it was composed—its spacious halls and polished oaken pannels—its furniture (a portion of which, and of the family plate, though rich and expensive, was of the fashion of other days), and in its numerous and well-appointed retinue of household attendants, there was something of Baronial grandeur. For more than one generation it had been known as the abode of opulence, refinement, and hospitality. It stood upon one of those elevated bluffs which are much prized in that champagne country, and presented opposite fronts, which were ornamented with spacious Grecian porticos. One of them looked out upon a grassy lawn of eighty or a hundred acres, decorated with stately oaks, apparently almost coeval with the alluvial soil in which they had vegetated. On the right were gardens, in which were domesticated many of the flowers and fruits, and culinary productions of northern and tropical climates.

Thomas Lynch, Jr., studied law in England before returning to South Carolina in 1772, but decided not to practice. After his father gave him Peachtree, he married Elizabeth Shubrick and began his short career as planter and active public servant. Upon his death, the property passed to his sister who married a Scottish adventurer, John Bowman.

Bowman was something of an eccentric. It was he who hired Jonathan Lucas to build the first rice mill in South Carolina. He is also said to have built a wind-powered sawmill on nearby Cape Romain, where he kept a summer house floating on a barge. If a hurricane struck suddenly, he expected to be blown to safety.

Archibald Rutledge tells a romantic tale, that appears to have some basis in fact, of Bowman's daughter. The young girl died of fever, and the heartbroken father placed her coffin above ground and piled the dirt about it until he had built a small mound. A large pine grew from the center of this grave, which could still be seen until fairly recently.

St. James Santee Church

Built ca. 1768, St. James Santee Parish, Charleston County, National Register, National Historic Landmark.

The fourth church to serve St. James Santee Parish, this lovely building was erected in 1768. By that time the original Huguenot congregation of 1706 had become thoroughly Anglicized, and the upper part of the parish, "English Santee," had been cut off and given a new church at St. Stephens. After deciding that the Echaw Chapel they were using was inconvenient, the congrega-

The church stands today in a remote, almost abandoned stretch of wilderness. Although once said to be a crossroads, no evidence of this remains and there is only Brick Church to testify to the existence of a prosperous rice growing community on the South Santee River.

tion applied, under the direction of Thomas Lynch, Jacob Motte, and others, for another church to be located near the "Wambaw Bridge."

This resulting building was apparently influenced by the sophistication of the newly completed St. Michaels in Charleston. It is a surprising departure from the rural church architecture of its day, for attached to each face of the structure are classical porticos supported by four brick columns, complete with molded bases and capitals.

The interior has what had become the conventional floor plan, with high backed boxed pews separated by a cross axis of tile paved corridors. During the eighteenth century the pews were rearranged and the chancel may have been moved to the east wall. The rear portico had been enclosed as a vestry room. Placed before the Palladian window of the north wall is a finely crafted pulpit, the work of one of the church's modern-day priests.

With the coming of the Revolution, rectors of Anglican churches faced the same choice as their parishioners, and the choices they made were not always easy ones. Of particular interest is the decision of the Rev. Samuel Fenner Warren. A native of Suffolk, England, he arrived here in 1758 and was immediately elected Rector of St. James Santee, a position he held until 1774. When he went back to England on a visit he was pressed to remain, but felt that his affections and loyalties were for America. He returned in the midst of the war and resumed his parish functions. His son, Colonel Samuel Warren, distinguished himself in battles of the Revolution, and a barely legible record of these exploits remains on his tombstone which has been moved to beside the church.

Never painted, these beautiful box pews date from the eighteenth century. The communion silver was stolen during the War Between the States but was later returned.

Only a single room wide, the long wings of Harrietta's unique floorplan guaranteed good ventilation. The probable designer was Thomas Pinckney, who once ran for the Vice Presidency on the same ticket with another amateur architect, Thomas Jefferson.

Harrietta

Built ca. 1800, Santee River, St. James Santee Parish, Charleston County, National Register.

King Jeremy of the Sewee Indians is the first recorded resident of the Harrietta site, but by 1709 the entire south side of the Santee River up to Jamestown had been granted to settlers by the Proprietors. This section probably went to John Fenwick. In 1735, Gilson Clapp, a wealthy Dorchester merchant, received a royal grant and built a house here at "Washaw" for his wife Margaret, a daughter of Thomas Lynch, Sr. A dizzying series of marriages and deaths followed over the next half-century; of this couple's four children, only Mary lived long enough to assume her inheritance. She and her second husband David Deas had three girls who inherited part interests in the land, and it was Alexander Inglis, the widowed husband of their daughter Mary Deas, who finally received the Harrietta portion. In 1785, he left it to his son David, who conveyed it in 1791 to John Wagner, who sold it to Mrs. Harriott Pinckney Horry of Hampton in 1806.

This front parlor of Harrietta is finely furnished with mahogany Chippendale china breakfront, eighteenth-century tea table and Philadelphia-made Queen Anne chair.

Harriott meant to give the plantation as a gift to her daughter Harriott, who had married Frederick Rutledge in 1797, and so she began the present house. In the meantime, her only son Daniel, after changing his name to Charles Lucas Pinckney Horry and marrying a niece of Lafayette, decided to live in France. That left the Rutledges living at Hampton, and construction was halted. About 1828, work was resumed on the interior, the intended resident this time being the Rutledges' sec-

Following the Revolution, an interior this elaborate was more often found in a townhouse. In fact, Charleston's Nathaniel Russell house, also tentatively credited to Pinckney, has strikingly similar woodwork. The hand-painted wallpaper dates from the Shonnards' restoration and the table is by Charleston cabinet maker, Thomas Elfe.

ond son Frederick. However, when the oldest son chose a career in the Navy, Frederick, Sr., remained on at Hampton, and construction was halted again. In 1858, the house was bought and finally occupied by Stephen D. Doar.

At his death in 1872 it passed to Stephen Doar's son David, who was one of the last in the state to continue planting rice in commercial quantities. At his death in 1928 the house was sold, with some rooms still unplastered, to the Sonnards who restored it completely, adding new entrance steps and a stairway inside.

As would be expected of a building under construction for the entire first half of the nineteenth century, Harrietta has elements of both the Federal and Greek Revival styles. Of particular interest are the unusual double front doors. We will see this arrangement repeated often in later houses, but here a third false door is placed between to present a balanced entry. Instead of a standard central hall, the front of the house has a pair of rooms adjacent to one another. Behind these two rooms, a cross-axial hall connects the rooms of the main house with rooms in attached wings. Also in 1797, Harriott Horry's brother, Thomas Pinckney of Fairfield, had begun to build a house at nearby El Dorado. There is enough similarity between the two houses to suggest that he may have been the architect of Harrietta as well.

Tibwin

Built ca. 1805, Tibwin Creek, St. James Santee, Charleston County.

John Collins received a grant for the Tibwin property in 1705, and it remained in the Collins family until purchased by William Matthews in 1794. Matthews is thought to have built the house, originally a story-and-a-half cottage, in 1805.

In September of 1822, a devastating hurricane came ashore here at Cape Romain. The destruction at the nearby Santee Delta was great and many lives were lost. As a result large conical storm towers were built to protect the slave population in the future. In addition, the summer colony of the planters was moved from Cedar Island in the mouth of the Santee River to the site of present-day McClellanville. Mr. Matthews may have thought this a curious choice for Tibwin was near the site of this new safer settlement, and the Charleston *Courier* reported that the same storm had destroyed both his dwellings and outhouses and seriously injured his crops for a total loss of about $7,000.00.

The house was probably not a complete loss, for both tradition and the evidence of the building itself suggest that it was actually moved back from the water's edge and rebuilt at this present location.

At the turn of the century, a second story was added, giving the house a new double portico façade. The first floor still contains the original woodwork. Of particular interest is the Adam mantel and gougework cornice of the front parlor.

An early rice mill located on the property was purchased by Henry Ford during the Great Depression and is now on display at his museum in Dearborn, Michigan.

Wedge

Built ca. 1826, Santee River, St. James Santee Parish, Charleston County, National Register.

Born in Cumberland, England in 1754 and educated as a millwright, Jonathan Lucas sailed for America shortly after the Revolution. Bad weather forced his ship aground at the mouth of the Santee River, and coming ashore, he saw that the laborious mortar and pestle method was being used to clean the hulls from rice. At nearby Peachtree he built the first of his famous waterpowered pounding mills and revolutionized the processing of rice. His son William, after an apprenticeship in the account houses of Charleston, joined his father in business. He prospered at this, and eventually acquired a townhouse in Charleston, a summer home in Aiken, a 4,000-acre estate on Murphy's Island, and some 500 slaves.

To keep an eye on his Santee properties, William and his wife Charlotte Hume of Hopsewee built this fine Greek Revival home on a wedge-shaped piece of property. The building is of a traditional design, but is exceptionally well proportioned. A

William Lucas' Wedge boasts this exceptionally well-balanced Greek Revival facade.

Prior to state ownership, the dining room at the Wedge appeared as seen here. The interior trim, in keeping with the Greek Revival style, is comparatively simple.

pair of stone steps lead to a flat-roofed portico that is supported by finely "subdued" Doric columns and trimmed above with unusual wooden urns. The garden still contains the many magnolias and camellias that were placed there by the builder, and though the original English-style gatehouse is gone, the fleur-de-lis gate post of the entry remains.

Lucas' neighbors could not spend more than seven months in the countryside without risking death from the fever, and he was noted by them as staying considerably less than that. Ironically, the Wedge today is the home of a state-sponsored center for the study of vector-borne diseases. With the malaria-carrying mosquito becoming resistant to pesticides and the malaria parasite to medication, here, once again, science is addressing the age-old problem.

David Doar of Harrietta and others dated the increasing death rate in the years following the Revolution to the impoundment of waters, which indeed was providing more breeding grounds for the insect as more and more rice fields were planted. No doubt this contributed to the problems, but modern theory indicates it is possible that a more virulent strain of the disease arrived from Africa at about that time.

Millbrook (Annandale)

Built ca. 1833, Santee River, Prince George Winyah Parish, Georgetown County, National Register.

Millbrook Plantation was part of a tract granted to the first Thomas Lynch in 1731, and was acquired by Andrew Johnson shortly after the Revolution. Johnson had Jonathan Lucas build the first tide-operated rice mill there and passed it on to his son William, several of whose children planted in the area. Of these the most exceptional was An-

Robert Mills did not design Millbrook, but he is responsible for similar columns that front the nearby Georgetown Courthouse. Though common in the South, Greek Revival porticos of this proportion were rare in our Low Country.

drew, who after a visit to his homeland in Scotland, changed the name of the place to Annandale and the spelling of his own name to Johnstone. He built the present house in the 1830's.

Andrew appears in the 1850 census growing 900,000 pounds of rice, which probably made him the most successful planter of the area. In an assessment for a proposed sale a year later he valued his property at the substantial sum of $220,000. This would have included half of his 400 Negroes, "a remarkably prime gang," 1,100 acres of rice lands, 1,200 acres of high land, his thrashing mill, "the furniture in his house (with the exception of a few favorite articles of little value)," his "flats and boats—wagons and carts—mules and horses—stock cattle and oxen—timberland," plantation tools and a house on South Island, six carpenters (or perhaps more), and a first-rate engineer who was also a blacksmith and bricklayer.

The sale did not go through and the family apparently continued to prosper. According to the 1860 census, Johnstone's slaves grew a staggering two million pounds of rice.

Without the usual basement and minus the rear addition of the 1880's, the house was not large. Considering the wealth at his disposal, Andrew's home hardly appears extravagant. Clearly in the Greek Revival tradition, the massive columns of its portico are almost too grand for the building they adorn. Inside, the attenuated columns of the well-preserved mantels and molded window frames with their carved acanthus leaves appear more familiarly proportioned. The entrance leads through a pair of doors, whose counterparts could be found directly across the delta at Harrietta.

In the midst of the War Between the States, Millbrook was sold to George Alfred Trenholm, who would become Confederate Secretary of Treasury a year later. The job won him few thanks and led to his eventual arrest and imprisonment. The property went to his son-in-law William Hazzard and then to the Santee Rice Company. At the turn of the century, rice planting was passing, but Millbrook, along with other delta properties won a new and well-deserved reputation for its duck shooting.

Pine Grove

(Physician's House of Millbrook)

Built ca. 1834, Santee River, Prince George Winyah Parish, Georgetown County.

Just behind the Millbrook house is a small pond and beyond that, in a second grove of oaks, is the plantation physician's house. Designated today as a separate plantation, the Pine Grove house is said to date from the construction of the main house in 1834. If so, it is an early example of the Gothic Revival cottage, for it would be another decade before plan books made the picturesque "country homes" a familiar sight in America.

This Gothic Revival cottage at Pine Grove was originally built as a home for the Millbrook plantation physician.

Original to the building is the central section. Inside, high ceilings with deep plaster cornices distinguish the simple but large rooms.

The Pine Grove of today is unique, for it boasts the only beaver colony we will see. Alligators usually get the better of these holdovers from Colonial Carolina, but here the beavers are holding their own.

Laurel Hill

Built ca. 1852, Sewee Bay, St. James Santee Parish, Charleston County.

The simple and harmonious exterior of this story-and-a-half raised cottage has remained essentially unchanged since its construction by Richard Tillia Morrison in 1852. The plantation logbook of that year lists expenditures for Charleston carpenters and materials. The records of the next decade include accounts of rice, cotton, and large amounts of forest products—cord wood, lumber, shingles, turpentine, resin—transported from Laurel Hill.

"Richard Le Tellier" was the name of the ship that brought Scotch refugees to America following the Battle of Culloden. It took several generations, however, for what W. J. Cash called this "pioneer stock" to make its way down the eastern seaboard from Pennsylvania to Charleston. Even then they didn't stop, for in the early 1830's Morrison and his father went on to Alabama and settled there, only to be driven out by an Indian uprising. Returning to Christ Church Parish, Morrison worked for some time, apparently as an overseer, and then purchased Laurel Hill in 1850. Within a few years he accumulated over 7,000 acres, and enjoyed a prosperity cut short by the War.

Married twice, first to Eliza Venning and then to Eliza Toomer, Richard T. Morrison had 16 children and lived to be 94. His epitaph begins "The union of acute intellect, restless nervous energy, and powerful physique which delights in taking arms against a sea of trouble and by opposing, ending them." It then goes on to list his success at draining swamplands, his work as an engineer in the War Between the States, his founding of McClellanville, four miles north of Laurel Hill, and his attempts to bring a railroad to the area.

Laurel Hill remained in the family as a working farm until the 1950's. Though the land continued to be planted, the house was abandoned at that time. Encroached upon by the highway, it seemed almost beyond repair, when descendants moved the derelict structure to a new site and renovated it.

Doors, mantel, newel post, all are original to the house. A single board molding which runs beneath the ceiling is the only embellishment in this plain but serviceable Scotch Presbyterian home.

Recently returned to family ownership and refurbished, the Laurel Hill house was moved over a mile to this made-to-order site bordering the marshes of Bull's Bay. Sixteen children were reared here and innumerable descendants followed.

"The family of the Gaillards lie here interred." Rev. Dalcho wrote further of St. Stephen's, "this handsome country church would be no mean ornament to Charleston."

Eutawville

Considered high ground and thus unsuited for rice, the area around Eutawville was only sparsely settled before the Revolution. The presence of the heavily traveled Nelson Ferry road, and the homes of both Francis Marion and Gen. Moultrie, made it the scene of heavy and destructive fighting. Following Independence, the loss of the indigo bounty and the freshets along the upper river drove planters further into debt, but prosperity was not long out of reach. In 1792, Whitney's gin made short-staple cotton suddenly very profitable and the construction of the Santee Canal in 1800 provided easy access to the Charleston market. By the turn of the century Eutawville planters were building homes that reflected this prosperity.

St. Stephen's Parish Church

Built ca. 1764, St. Stephen's Parish, Berkeley County, National Register, National Historic Landmark.

In May of 1764, the vestry of the recently organized St. Stephen's Parish met to arrange for the construction of their new church. One member, indicating perhaps a degree of foresight, declined to serve as a commissioner. Another vestry member agreed to supply 150,000 "good merchantable bricks" at eight pence per thousand, and a second member was engaged to saw "all the timber Plank & Boards and the Laths to be wanted for Building . . . Plates and all below to be cypress except the floor and all above the plates to be of pine." A year later, the commissioners for building the church turned down the bricks as "Entirely too bad," and the lumber dealer in the vestry agreed to provide bricks subject to the approval of the Commission and fired to "the size of the moles to be equall in Bigness to Mr. Zackry Villepontoux's." Monies began to be disbursed to William Axson and Francis Villepontoux. Both left their signatures on the church, and together they received slightly over 1,000 pounds to cover the work.

When the church was completed, Francis Villepontoux joined the vestry. Five months later three commissioners resigned over the arrangement of the pews, which was said to be "of a very singular Nature." Apparently this was a reference not to the assignment or sale of pews but to the actual arrangement, for a year later a pew was sold on the condition that the owner "shall not by any manner of means or ways alter the present uniformity of the sd pew, as it is now built, by raising and the same."

Despite—or perhaps because of—disagreement over how the church should be designed and constructed, it was solidly built. The vestry had discharged its responsibility well. The design is quite unusual. In order to accommodate a coved ceiling similar to one in the newly completed St. Michael's, a high gambrel roof was built. The ends of this roof have curvilinear gables suggestive of an earlier era. A common complaint is that the roof appears too massive and the Palladian window above the altar too small, but the end result is pleasing enough, and the interior is impressive. Dalcho called it "one of the handsomest country churches in South Carolina," which is no doubt what the commissioners and builders had intended.

Unfortunately, the St. Stephen's building was to have only a brief period of use. Anglican enthusiasm waned after the Revolution. A year of flooding left much of the area unfarmed and many plantations permanently deserted. The Chapel at Pineville gradually became the center of worship for the shifting population, and in 1819 Dalcho wrote also that "the noble edifice is falling into ruins." It was well into this century before regularly held services began again.

Capt. Peter Gaillard was fleeing from the waters of the Santee when he settled here at the Rocks. Ironically, the Santee-Cooper project of 1940 caused the waters to rise permanently and the house had to be moved to higher ground.

The Rocks

Built ca. 1805, St. John's Parish Berkeley, Orangeburg County, National Register.

Three generations removed from the Jamestown settlers of 1695, Capt. Peter Gaillard was in many ways typical of the other Huguenots who would follow him into upper St. John's Parish. He had married Elizabeth Porcher, who would bear him 12 children, and had begun farming a St. Stephen's plantation given to him by his father Theodore. The Revolution came and, after the death of his loyalist father, he joined Marion's troops and fought with "great gallantry." Following Independence he was forced into bankruptcy by the fallen price of indigo. Travelling to Charleston, he turned all of his property over to his creditors, but the slaves were kindly returned to him, and with the proceeds of a lottery ticket, he purchased the first tract of the Rocks with the hope of feeding his family. His first cotton crop, planted in 1795, brought the beginnings of a fortune and a "reputation as a planter that was immense."

Capt. Gaillard kept a detailed record of his house construction. On September 1, 1803, they began to make bricks. Soon cypress was split for shingles, sashes built, and bricks laid. By April 7, 1804, the house was framed. On July 3, he "agreed with Mr. Walker to have my chimney pieces and doors made at north . . . The doors to be double worked, two chimney pieces to be done in a genteel but plain style, and five others being for bedrooms to be very plain." July 9 the second coat of paint went on, and the house was finished and inhabited in time for his oldest daughter's wedding.

Leize Palmer Gaillard left behind a short but detailed account of the Rocks and its residents. She grew up there following the War and tells of lancing tournaments and dances in a battered, but certainly unvanquished South. This excerpt describing Christmas dinner is just a sample, a taste if you will, of the whole:

> The dinner was always the same. At one end of the table, a turkey stuffed with spinach (the housegirl or cook plucked a huge dishpan of spinach and washed it thoroughly, then as much butter as was conveniently spared, probably a half pound or maybe a little more was put in a big frying pan on the stove, the spinach dumped into it and as many eggs as you could muster stirred into that, and then stirred and stirred until the mass of spinach and egg was a gigantic pile of green scrambled egg, and you stuffed the turkey). At the other end of the table another turkey stuffed with the more usual bread crumbs, egg, etc. At various places along the table were a boiled ham, a large one, a huge roast of mutton, leg of lamb, from four to six boiled chickens, big ones with a very rich sauce with hard boiled eggs stirred up in it, dishes piled with snowy rice. I asked Mary once how much rice did she have cooked for that dinner, and she said a peck of raw rice, and a negro from the negro quarters cooked it out in the yard in an iron wash pot. In addition there were big pans of macaroni, sweet potatoes, Irish potatoes, creamed artichokes sometimes, glass dishes of whole artichoke pickles, a decanter of whiskey and two or three decanters of wine. Then for dessert, Charlotte Russe, wine jelly and Syllabub and *always four kinds* of pie, cocoanut, lemon, mince, and sweet potato. I don't remember whether there was cake or not, perhaps I was just too full to be impressed by it.

Loch Dhu

Built ca. 1812, St. John's Parish, Berkeley County, National Register.

Loch Dhu takes its name from the Gaelic "black lake," a description the Scotch owner gave to a small dark pool on the property. William Kirk received a grant for the land in 1749, and his son Gideon appeared as a resident planter of the par-

Typical of upper St. John's is this house of the black lake, Loch Dhu.

ish in the 1790 census. The house was built by grandson Robert in 1812, a date recently found carved into one of the chimney bricks. An inventory of Robert's estate in 1829 suggested a modest wealth, which his son Philip would maintain. Philip served in the South Carolina House of Representatives before the War. He married Gabriella, the youngest of Francis Marion's daughters, who died soon after.

The box-like house with steep hipped roof and tall chimneys is typical of the area, as are the pair of entry doors beneath the front piazza. Each door opens into a separate front room, which opens into the other. Both rooms open onto a stairhall in the rear, which, in turn, gives access to two small back rooms. By eliminating the front hall, the front rooms were enlarged and, as Gene Waddell points out, they could be used for entertaining much as the double parlor was. Certainly the layout must have proved functional, for it was used too consistently to be only the affectation of a neighborhood. Also found in this area was a curious small room that opened onto the stair landing halfway between the two floors. Elsewhere called the library, the one at Loch Dhu is now a bath. The second floor has four bedrooms divided by a central hall. With the addition of a cellar and occasionally some rear wings, this floor plan was repeated often in Upper St. John's Parish.

The woodwork of the interior is fairly restrained, at least by the neighborhood's standards, but the mantels and corners of the front rooms are decorated with gouged festoons, sunbursts, and rope motifs. The woodwork of the living room has been repainted, but the dining room retains panels of wainscoting stained and grained to simulate a mahogany trim and light wood center, with doors lightened, and baseboards black.

Walnut Grove

Built ca. 1818, St. John's Parish Berkeley, Orangeburg County.

Here is the house that James, the son of Capt. Peter Gaillard, built for himself in 1818. The double doors were replaced about 1900 with a single entry opening on a central hall. A metal ceiling was hung in the parlor and Victorian modernization given throughout, but the house still retains some features missing from the Rocks, his parents' house. Full wings come off each side, and below is an eight-foot basement. Though there was an outside kitchen, a "warming" kitchen was set up in the basement as well as cool storage for daily products and other perishables. Like most houses of the area, the piazza entrance was oriented to the south. In front of the entrance remains a garden, and the imprint of a circular drive that brought guests to the house's granite stepping stone. Some nineteenth century outbuildings survive, and on every side are large magnolias and oaks.

Inside the house are mantels trimmed with Philadelphia marble, elaborate overmantels, and wainscoting. The upstairs still shows evidence of early polychromatic staining that refuses to be hidden by subsequent paint. A narrow stair leads to the attic, which contains a trap door onto the steep roof. This fairly common feature provided easy access to

By 1818 the importance of gougework in mantels and cornices had increased, and the overmantel at Walnut Grove is fairly detailed. The pressed tin ceiling is a later addition.

A Victorian restoration removed Walnut Grove's double doors, and the wooden steps were replaced with a flared brick entry. The current owner reminds us that, although spacious, the house was not considered large. Seven Gaillard children and 11 of the Connor family occupied it in turn. Such families were not uncommon and, with resident relatives and servants, went a long way in filling up a house.

repair cypress shingles or to put out a fire in a chimney or on a burning roof.

James Gaillard was too old to serve in the War Between the States, so he helped the cause by making his home a refuge for the wives and sisters of soldiers. General Hartwell and his soldiers looted the house, ripping off doors and shutters, but when they threatened to burn the house the women positioned themselves strategically throughout and refused to leave. The house was saved, but many valuables were lost, among them "nine dozen beautiful linen damask napkins."

Lawson's Pond

Built ca. 1823, St. John's Parish, Berkeley County, National Register.

In a sense, Lawson's Pond is the centerpiece of its community. Writing before the Santee-Cooper project, Thomas Waterman called it "the only house in the area which approached the Rocks." Since the time of this estimation the Rocks was moved, sacrificing, it can be argued, at least a degree of its authenticity. Lawson's Pond, however, appears almost untouched. Not even a single coat of paint has been applied to the cypress exterior.

The details of the interior are elaborate. Each

A spring morning at romantic Lawson's Pond.

Gougework unrivaled in quality appears in the trim of Lawson's Pond.

room of the main floor has its own design. Mantels and doors and window frames are heavily carved with sunbursts, fans, Chippendale-style fretwork, and variations of the rope motif. The wainscoting, doors, and stairways are marbleized.

When Charles Cordes Porcher inherited Lawson's Pond from his father in 1818, its slaves, livestock, and land were valued at over $42,000. Charles is thought to have begun the house in 1823 in preparation for his marriage to Rebecca, a daughter of General Francis Marion. Obviously he continued to prosper in the years after, for the 1860 census shows he owned 1,000 acres of improved farmland and 110 slaves, and produced 150 bales of cotton. Despite his prosperity, he never completed his house.

The craftsman responsible for Lawson's Pond is assumed to be George Chaplin, the same "Yankee carpenter who astonished the people by his skill" on neighboring Springfield. The plantation book there shows that owner Joseph Palmer paid Chaplin a salary of $60 a month over a period of seven months and three days, and that all work was apparently finished satisfactorily.

Here on Lawson's Pond, however, work came to a halt just short of completion. Two upstairs bedrooms were left unfinished. The entablature above the windows was omitted, and none of the exterior woodwork was painted. Perhaps a temporary lack of funds caused him to stop, or maybe Chaplin had returned north. Frederick Augustus Porcher's memoirs, however, may offer a further explanation. All of General Marion's daughters were famed for their charm and beauty. "Oh what a luxury it was to dance the six-handed reel with one," he writes, "and with two other sisters to make up a set." All but one, unfortunately, died tragically young. We know that Charles' wife, Rebecca Marion, died in childbirth in 1827. Perhaps his incentive was lost at this point, for the heirless owner lived for another half century in the unfinished house. This romantic mystery is further embellished by the recent discovery of ancient carpenter's tools in the attic.

Numertia

Built ca. 1840, St. John's Parish, Orangeburg County, National Register.

Numertia takes its name from the Latin "Numerus Tertius" for it was constructed on property designated as "tract three." Major Samuel Porcher built the house for his grandson Richard S. Porcher, who lived there a short time with his wife before moving to Pendleton. In 1856, he sold it to his newly married cousins Lydia Catherine and Christopher Gaillard.

Not typical of the other houses built by these families, Numertia has a single entry and a central hall, which makes it wider, and a gabled roof. The siding is simple square-dressed lumber cut on the property, it is said, by whipsaw. The porch columns are slender, square, and tapered. The interior is simple as well. Hand-planed wainscoting runs throughout, but doorways and mantels, while substantial, are not elaborate.

Obviously the emphasis here was on solid construction and functional living space. The well-preserved house has its full complement of rooms and more. There is a great basement complete with warming kitchen. Several outbuildings have survived, as well as a handsome planting of magnolias.

In 1911, Christopher's grandson William S. Gaillard returned, bought the divided land back and, with the help of Gabrielle Marion Kirk, whom he married in 1916, started a prosperous dairy that continued operating until the 1970's. During the 1930's the family reestablished a tradition of fine horsemanship that in the Eutawville area dated back to the Revolution.

Solidly constructed and commodious, Numertia was built for functional living.

This unusual building at St. Julien is the only Italianate plantation home in the survey.

St. Julien Plantation

Built ca. 1850, St. John's Parish, Orangeburg County, National Register.

Most of upper St. John's Parish, including those plantations seen so far, was included in the original 48,000 acres of Raphoe Barony, granted to John Bailey in 1698. The land was not settled then and may not have even been surveyed, for it was declared vacant in the mid-1700's. Other grants had already been made in the surrounding area—one to Joseph de St. Julien in 1736. This property was then bought by the Gaillard family and passed through inheritance to Thomas Porcher, who built the house and outbuildings in the early 1850's for his son Julius and new daughter-in-law Mary Wickham Porcher.

Julius had been educated in medicine, studying in England and at South Carolina College. He entered the Confederate service and rose to the rank of lieutenant colonel before being killed in the Battle of Missionary Ridge. After his death, the plantation continued under the management of Frederick Connor and in 1906 was bought by Connor's descendants.

Dating from the time of construction are an avenue of oaks planted in a serpentine J in honor of the property's last name and a garden of ancient japonicas. The outbuildings, a stable trimmed with lattice, and a split-pole warehouse are also original.

The interior of St. Julien has apparently been remodeled during this century, but the fine marble mantels of each room are original and much of the rest of the interior trim is as well. The bell-cast shed roof of the front porch, the wide eaves with decorative brackets and the two-story L with a single-story wing are clear expressions of Italianate style. Certainly early for the South, this type of building relied on the Italian Renaissance villa for its inspiration, utilizing a style that was contesting the rising surpremacy of the Gothic Revival. Charleston architects E. B. White and Jones and Lee were designing buildings in both styles, but with the coming of war, few of either were to be constructed as plantation homes.

Edisto River System

EDISTO ISLAND

THE little island of Edisto enjoyed an isolation, wealth, and secessionist fervor that earned it the title of "the royal principality of Edisto." Settled early by Scotch and Welsh immigrants, attempts to grow rice on the island were abandoned quickly in favor of indigo. It was, however, the planting of Sea Island cotton around 1796 that brought great profits.

In an 1808 appendix to Ramsay's history, Edisto is described in detail. There were 236 white inhabitants and 2,609 slaves living on the 28,811-acre island, and the annual income from cotton was $321,300. This amount evenly divided would have converted into the comfortable sum of $8,683 for each married white couple.

Land was already at a premium, and by mid-century virtually all available acres would be in cultivation. Because of the narrow confines of the island, residents were noted to prefer their elegant and swift sailboats to useless thoroughbred horses. And since Edisto afforded little habitat for game, local sportsmen contented themselves with "a great variety of excellent fish."

Ramsay's history describes the climate as particularly sickly, with fever prevalent. Over the 15 years before the history's publication, some families had disappeared entirely and "a number greater than three-fourths of the inhabitants had died."

Edisto planters paid great attention to education, and nine boys of the tiny population were reported away at school, four of them enrolled at Princeton or Yale. The islanders boasted that in their enthusiasm for religion and their funding of churches they could not be rivaled. "Dancing parties are confined to the temperate seasons of the year. They are neither so frequent nor so eagerly pursued as they are reported to be in other parts of the State."

Edisto's pious and industrious planters enjoyed increasing profits until the War, but Union forces occupied the island early, and Sherman settled freedmen and women there. Most plantation houses escaped destruction, and the planters eventually regained their property, but the day of island royalty had passed.

Brick House

Built ca. 1725, Russell Creek, St. John's Colleton Parish, Charleston County, National Register, National Historic Landmark.

In 1860 Col. Joseph Evans Jenkins stood at a secession meeting and announced, "Gentlemen, if South Carolina does not secede from the Union, Edisto Island will." His home was Brick House, the oldest on the island. It had been built by Paul Hamilton around 1720 and had been bought by Joseph's father "because he was tired of pirates worrying him" on the more exposed end of Edisto.

Though the house burned in 1929, it is documented in drawings and photographs, and much of the walls still stand. The building measured 40' × 36' and was two stories high. The ground floorplan was similar to that of Mulberry, with an entrance into the main drawing room. A heavy emphasis was placed on brickwork and accompanying trim. The corners of the building and window

Though the inspiration for Brick House was probably Dutch or English, it once resembled a French chateau. The builder Paul Hamilton was not Huguenot but Presbyterian.

jambs were trimmed by stucco quoins to look like stone, and stucco bibs and flat-keyed arches touch the stucco belt that circles the building at the second-story level. On the sides were stucco panels, each with a blind arch below and a false window above. While much of this decoration remains, it takes some imagination to picture the high "bell-cast" hipped roof, the flanking buildings, and the richly carved and painted woodwork of the interior.

As to the builder, we have this testament to his nature written by his nephew:

> Paul, my father's oldest brother, built a mansion on Edisto Island, the brick of which he imported from Boston in New England (the sand and gravel mixed in the mortar was all brought from Pon Pon River that it might be free from salt), and the timbers of which were all cypress cut from his own lands on Pon Pon River. The inner work was of cedar, and he allowed no wood to be used that had not been housed and seasoned seven years.

Old House

Built ca. 1750, St. John's Colleton Parish, Charleston County, National Register.

Originally known as "Four Chimneys," Old House gained its present name when Brick House burned and it became the oldest dwelling still intact on the island. Built between 1735 and 1760 by William Jenkins, the house was sold before the Revolution to the Seabrooks, who carried out extensive renovation early in the 19th century. The bulk of this work was confined to the exterior. A large portico with Tuscan columns, dormers, Palladian entrance, keystone lintels, and fan-shaped windows both front and back gave the house a beautiful balanced facelift. Judging by the solid construction of the foundation and joists, Jenkins had built well to begin with, and the interior was

Refurbished by the Seabrook family early in the nineteenth century, Old House emerged with a new, well-proportioned exterior.

Most of the interior of Old House is thought to be original. Of particular interest is the high wainscoting in the entry hall.

only altered by the addition of a keyed archway in the hall and by some minor trim. Apparently original to the house are the unusually high wainscoting with raised wooden strips in a crenellated pattern, and the simple one-piece mantels with raised diamond design.

Seaside

Built ca. 1802, Scotts Creek, Edisto Island, St. John's Colleton Parish, Charleston County, National Register.

Once called Locksley Hall, the house at Seaside stands just out of sight of the ocean in that pirate-infested vicinity which Joseph Jenkins deserted for Brick House. The property was bought in 1802 by William Edings, and he is assumed to have built the house shortly after.

Distinguished as the only Federal-style brick house remaining on Edisto Island, the two-story stucco building is similar in plan to the Charleston

Marble trim, fluted pilasters on the mantel, paneled overmantel, and a cornice with gougework and dentil trim—these are some of Seaside's "modest Federal embellishments."

"single" house, one room deep with a piazza across the southern exposure. Several additions have been made and the piazza lowered, but the interior still features the restrained decorations of the original: dentil-work cornices, wainscoting, and finely worked mantels.

Edings was a planter of Sea Island cotton, but as indicated by his relatively modest house, he did not enjoy the extravagant success of some of his neighbors. He served in the state legislature from 1856-1857 and died the following year.

Also owned by the Edings family was the nearby summer resort of Edingsville (also called "The Bay"). Lots were leased to other families and, by 1820, 60 large homes had been built there, as well as two churches and an academy. "The Bay" offered the planters a refuge from the country fever, which on Edisto Island took a particularly heavy toll. From this village they could ride each day to inspect their plantings and spend their leisure in genteel company. As with several other oceanfront communities, all reminders of "The Bay" have washed into the Atlantic.

William Seabrook House

Built ca. 1810, Steamboat Creek, Edisto Island, St. John's Colleton Parish, Charleston County, National Register.

Construction of William Seabrook's house in 1810 distinctly affected the island's subsequent architecture. Tradition claims that Hoban, who designed the White House in Washington, D.C., did this house as well. It is true that this architect practiced in Charleston for a short time during the 1790's and Seabrook was a boyhood friend of Hoban's understudy Robert Mills, but there is no proof of a connection. Still there is clear evidence of some guiding hand, at least to the extent we witnessed at Drayton Hall, and the massive portico and deeply carved wood and molded plasterwork of this earlier building invite comparison with the delicacy of work in the post-War era. The double portico, now familiar in Charleston and Beaufort, is handled particularly well, its upper story graced by shallow segmental arches. The interior shows an equal attention to craft, refinement, and balance.

A posthumous sketch in the *Southern Agriculturist*

William Seabrook's house, seen here from the landside entrance, was to become a model for other island homes.

Beneath a broad fanlight, triple doors open onto the hallway. The painting was removed from the house during the War Between the States, but a magazine article prompted its return many years later.

A formal garden is located between the house and nearby Steamboat Creek.

of February 1837 tells the story of the builder. Overcoming financial and educational disadvantages that resulted from the Revolution, Seabrook, at the age of 17, had taken over both his and his mother's estates. A pioneer in the use of salt marsh mud as fertilizer, he was one of the first to earn grand profits from Sea Island cotton. He established a steamboat line between Charleston and Savannah that stopped at the intermediate islands. For years he was an elder in the Presbyterian Church, and gave liberally to religious and educational institutions. He owned several large plantations as well as land scattered throughout the Sea Islands and built several handsome houses—this one being referred to as "his mansion house." He provided his older children with substantial estates and large sums of money, and left behind personal property valued at $376,916, a considerable legacy for the day. A second memorial remarks on his warm humor, generosity, and the fact that his death was mourned by everyone on the Island.

On the eve of the War Between the States the house was occupied by Seabrook's daughter, Josephine, and her husband, John Edings, Jr. Despite the bold secessionist declarations of the planters, the island fell quickly into Union hands. Soldiers quartered at Seabrook scratched on the walls threats of revenge against the Confederacy. It was well into this century before the house was reclaimed and restored.

This graceful double stairway is part of a modern renovation.

Middleton is a unique example of the Greek Revival style of architecture.

Middleton (Chisolm)

Built ca. 1830, Governor's Creek, Edisto Island, St. John's Colleton Parish, Charleston County, National Register.

Because of its easy access to Governor's Creek, this house has been called "The Launch," but at one time or another it has been known by the names of its resident families—Chisolm, Middleton, and Pope.

Although an 1821 survey shows a Chisolm house on the site, the present dwelling was probably built in 1830 by Oliver Henry Middleton and his new bride, Susan. He was the son of Governor Henry Middleton of Middleton Place, and she the daughter of Dr. Robert Trail Chisolm, from whom she had inherited the property.

The well-proportioned building at first glance seems deceptively plain, but there is no question that considerable thought went into its layout and exterior detail. The entire house is one room wide, guaranteeing good ventilation from the prevailing sea breeze, with porches at the front and back. The front entrance is dressed with the familiar fanlight and sidelights. The windows also appear to have sidelights, but this arrangement is a typical Greek Revival device for giving symmetry to an opening that is double the width of a regular window. These broad windows are trimmed with lintels and plain, heavy cornices that correspond well to the thick columns of the porches, the wide arcades of the basement, and the shallow hip roof.

The rooms have unusually high ceilings and excellent natural light. The interior is especially fine, with cornices, woodwork over doors, elaborately carved windows, marble mantels and a notable circular stairway.

To take advantage of the sea breeze, the second William Seabrook crossed the front and side of his house with a more livable piazza. The new owners are gradually restoring the gardens.

Oak Island

Built ca. 1830, Westbank Creek, Edisto Island, St. John's Colleton Parish, Charleston County.

Oak Island was built about 1830 by William Seabrook's son by his first marriage. This younger William had married Martha Edings, the sister of his father's second wife, and he built his house on a small oak-covered island a few miles from his paternal home. The kinship of Oak Island to the earlier Seabrook house is clearly evident, but here the ornamental double portico has been replaced with a broad functional piazza across the front, facing the water side of the house.

Inside, the interior reflects not so much a change of style as of proportion. Broader windows, higher ceilings, and taller doors recall the Olympian gran-

The stair at Oak Island is separated from the entry hall by a pair of handsome doors. The dark floors are indicative of the period.

deur of contemporary houses on the mainland.

Apparently even more attention was lavished on the grounds. A landscape gardener was summoned from England for the sole purpose of laying out the house's surroundings. Mrs. Seabrook left a description of the house that is well documented in photographs taken by the occupying Union army:

> I became so accustomed to its grandeur it ceased to inspire me. Federal troops landed at the wharf in numbers and came to this elegant old house with its twenty-one completely furnished rooms—lawns encircling the house occupied acres, outbuildings of every description, camellias of every known species, 1500 varieties of roses. A rustic bridge crossed to the Island. Walkways were covered with crushed shells. At the end of the avenue there was a park with many deer—including a white one. There was a quaint brick house where an iron chest of select wines were kept. Near the water was a dairy, a building made of crushed shells. Just beyond the dairy was a large long boat house. Sail and row boats were kept there and above were the bath houses. In the carriage house were seven or eight vehicles. The family used to ride to Virginia Springs and carriages were kept there during their stay and sent home the first of September when the family went to New York.

House builder William, incidentally, won a medal at the 1851 World's Fair in London for his cotton entry; that same year Dr. Heriot of Georgetown won for his rice. To put this achievement in perspective, Dr. George Rogers points out that at the same fair McCormick won a medal for his reaper, Colt for his revolving pistol, and Goodyear for India rubber.

Edisto Presbyterian Church

Built ca. 1830, Edisto Island, St. John's Colleton Parish, Charleston County, National Register.

The earliest settlers on Edisto Island were Welsh and Scottish immigrants, and the Edisto Presbyterian Church claims the distinction of being the oldest "existing in its original location and of unbroken continuity in South Carolina." Proposed as one of Rev. Stobo's churches, it had been established at least by 1710. At first the Presbyterian worshippers shared their building with the Baptists, and relied on circuit riders for their pastors,

Great wealth tempered by puritan austerity produced these pleasing results. Before the Revolution few congregations could compete with the official Anglican Church, but here in 1830 it was the Presbyterians who went unrivaled.

The pulpit has been lowered and the pews stripped to natural wood, but otherwise the interior of the Edisto Presbyterian Church remains unchanged.

but by 1722 they alone were using the church. Generous gifts from Paul Hamilton of Brick House and others during the next decade helped to secure the church's future, and its congregation continued to grow throughout the eighteenth century. In 1821, William States Lee began a memorable pastorate that lasted for 50 years. Sea Island cotton brought sudden prosperity to Edisto, and in 1830 the present building was constructed. The contract was let to Mr. Pillian, who received a $300 bonus from the oversubscribed capital building fund. In 1836 William Seabrook left the church $5,000 and at this time E. M. Curtis of Charleston replaced the small portico and installed a coved ceiling.

A large balcony with exterior entrances had been included for the slaves, whom Rev. Lee actively attended. From this period until the War, the Anglican and Presbyterian Churches were making earnest efforts in this direction, but could not compete with the Baptists and Methodists in attracting Negro churchgoers. Edisto was exceptional in that its church enrollment generally reflected the actual racial proportions of the population. During the War the building was used only by the freed slaves, but in 1866 Pastor Lee led his small congregation into a service and claimed the building "in the name of God and by the authority of the U. S. Government." Relations between the two groups did not remain strained; however, the black congregation was later given the Edingsville chapel for its use.

The graveyard that encircles the Church makes an interesting commentary on the self-contained island. Besides a monument to the founder, and fenced family plots, in one corner three abolitionist workers who drowned are fenced out. Behind another fence, a suicide who wasn't supposed to rest there is buried anyway. The loser of a duel is buried by the winner, his tombstone ominously proclaiming "prepare to meet your maker." In the midst of all is a tiny vestry building where the islanders could raise their voices without profaning the Church.

Presbyterian Parsonage - Edisto

Built ca. 1838, Store Creek, Edisto Island, St. John's Colleton Parish, Charleston County, National Register.

The site now occupied by the Parsonage was deeded to the Church in 1717 by Henry Bower, with the proviso that it was "for the benefit of a Presbyterian minister on Edisto Island." Other members contributed generously in the years that followed. By 1792, Pastor William Speers is listed as receiving £200, use of the parsonage, 40 acres of land, and a leave of absence during the summer. The original building burned soon after, and it was "resolved by the Corporation that the parsonage House to be built shall be in length 34 feet and in width 18 feet." Apparently this second structure was destroyed, but a portion of its red-brick basement forms the front foundation of the present building.

Simple both inside and out, the house commands a fine view of marsh and creek. In the attic where children once boarded we find that a pastor's son has left a eulogy for "Lady Bird who died June 13, 1917 . . . a gentle Horse, Good Traveller."

Cassina Point

Built ca. 1847, Edisto Island, St. John's Colleton Parish, Charleston County.

When Lafayette visited Charleston in 1825 he accepted William Seabrook's invitation to come to Edisto; while there he was asked to name his host's infant daughter. The French general chose the name Carolina Lafayette, and when grown the young woman and her sister went off to visit the Lafayette family and enjoy the hospitality of the "highest court circles."

Apparently Carolina preferred the social whirl of the North to life on Edisto. Her Philadelphian husband felt just the opposite, so for three years the lumber for Cassina Point lay on the ground while they debated. In 1847 construction finally began on a house much like that of nearby Oak Island.

Here, however, the exterior is simpler than at Oak Island, and the piazza is across only the front. The walls of the front hall have been replaced with twin sliding doors to enable the entire front of the house to be used as a single space. The interior is exceptionally well preserved—only one coat of paint has been applied since the original. The high baseboard is finely marbleized. In keeping with the custom of the times, the mantels are of real marble, and the large windows have the then-fashionable six-over-six panes.

During the War Between the States, the extensive basement was occupied by Massachusetts and New Hampshire soldiers whose still-legible graffiti include a half-dozen sailing ships and one steamer boiling smoke across the white plaster walls.

Bleak Hall Ice House

Built ca. 1850, Ocella Creek, Edisto Island, St. John's Colleton Parish, Charleston County, National Register.

Here on the ocean side of Edisto Island, John Townsend led a life of splendor that certainly rivaled that of the inland Seabrooks. A great mansion two-and-a-half stories high on a raised basement and a cupola at the top once towered above the surroundings, gardens, and outbuildings.

Townsend entered South Carolina College at age 15 and went on to Princeton. He returned to practice law in Charleston but soon came back to the island to handle the family's planting interest. He was elected to the state legislature before he reached the minimum age of 21, and later served in the Senate. He took an avid interest in public education and used his notable talents as an orator and author to urge on secession. In 1860 he went as a delegate to the Secession Convention. The agricultural census of that year shows him farming 1,731 acres in Sea Island cotton and owning 273 slaves.

Townsend's courtship of Mary Caroline Jenkins was a romance in the truest sense of the word. She was his "Lady Love," and he was her "Knight of the Golden Crest." After their marriage, she kept a light burning in the cupola to guide him home.

When Union troops were in control of the house, they used the same device as a navigational aid.

When Admiral Perry returned from Japan he brought with him a Japanese gardener named Oqui. Townsend is supposed to have gone to Washington and persuaded the gardener to come to Edisto, where he laid out an extensive and exotic garden. The garden boasted camphor, olive, and spice trees, oranges and all manner of vegetables and flowers. All is overgrown now except for the white poppies which surround the fine little ice house.

The steep roof gables trimmed with half-circle gingerbread, gable dormer, and fake Gothic doors and windows make this ice house an outstanding example of Gothic Revival architecture. Real doors and windows appear on the gable end. The holes in the flushboard surface were probably bored for ventilation. The building is pegged together, and the walls are filled with sawdust insulation. Inside, a partially below-grade construction of tabby with charcoal for further insulation kept ice that was brought in by ship.

EDISTO RIVER

Though easily linked to Edisto Island by waterways and settled by many of the same families, the plantations along the Edisto River were more likely to be in rice cultivation than the cotton that made Edisto Island famous. Willtown, an early political and economic center for these mainland planters, was replaced in later years by Adams Run, Walterboro, and, of course, the more distant Charleston.

Mount Hope

Built ca. 1807, Edisto River, St. Paul's Parish, Charleston County, National Regiser.

As first built in 1807 by Lewis Morris, Mt. Hope is a curious combination of the two architectural styles first seen at Fenwick Hall on Johns Island. Here executed in wood are octagonal "Federal" rooms placed at the center of a basic Georgian perimeter. This arrangement was particularly unusual, for the central room was encompassed by a second set of octagonal entries, but much of the original floorplan has been complicated by later additions and alterations.

The interior of Mt. Hope is simply executed, with only a chair mold and plaster walls. Ceiling cornices and mantels are in the traditional Federal style. Certainly one of the nicest features of the house is the tall windows that overlook the remnants of Confederate breastworks, and the beautiful Edisto River beyond.

Col. Lewis Morris, IV, who was the son of New Jersey's only signer of the Declaration of Independence, had come South with General Greene's army. After a courtship rich in romantic promise, he married Anne Barnet Elliott, and they built Mt. Hope on lot 16 of the Old Willtown site. They divided their attentions in the years after between his Southern home and Morrisiana, New York.

Mt. Hope's unusual central parlor is actually one polygonal room placed inside another.

Summit

Built ca. 1819, Swinton Creek, St. Paul's Parish, Charleston County, National Register.

The Swinton Creek site of the Summit house was originally granted to Landgrave Thomas Smith in 1694 and regranted to Joseph Blake six years later. John Bull, son of Stephen Bull of Ashley Hall, purchased it in 1730 and built a house and mill there. The house, judging by its ruins, was substantial enough to be one of Bull's chief residences. On his death it passed to Mary Middleton and her husband, Pierce Butler, a signer of the Constitution of the United States, who sold it soon after to John Dorsius. Further records have been destroyed, but judging by the configuration of the plats and family tradition, Daniel Jenkins bought it at a tax sale before his death in 1801.

William Wilkinson married Daniel's daughter,

When completed in 1819, this house marked the "summit" of mistress Amarintha's happiness. The wings were added recently.

Amarinthia, in 1813, and lived in the existing house until the estate was settled in 1816. Since the original house had fallen into disrepair and was known as "Rat Hall," it is little wonder, as the present owner points out, that the building of this new home would mark the "Summit" of Amarinthia's happiness.

Family account books list the cost of construction as of 1819, at $2,358.44, with $91 set aside for the brickmason's two chimneys, and $29.33 for the blue paint used in the interior. These costs were in keeping with Wilkinson's moderate success as a planter; he held other tracts of land, including large holdings in the nearby town of Adam's Run. Although rice was his principal crop, in 1829 he received a silver loving cup from the Agricultural Society for the best five acres of cotton. Wilkinson

The trim in Summit's Federal-style parlor was done with white pine instead of the usual cypress. Both tables were made by the owner. The lowboy was made in Charleston (ca. 1740).

was also an active member of the Willtown Presbyterian Church. At his death in 1847, the house, according to the terms of her marriage agreement, was left to his wife, but additional acreage seems to have been divided among his six children. The Wilkinsons, like others, suffered the privations of war and reconstruction, but the house remained in the family until the present owner bought it in 1966.

Summit was built in the basic farmhouse design of the day, with a two-story center balanced by shed rooms behind and a shed porch in front. Inside, the woodwork shows extensive detail of the Federal era, especially in the reeded pilasters and sunburst motifs of the mantels.

The house is curiously similar to Tombee on St. Helena. Both houses were built closer to the water than those on older sites, and though the configurations of rooms are different, the tabby foundations are almost identical. The interiors too, though separated by perhaps a quarter of a century, have mantels and other woodwork in common.

Moreover, both houses have benefited from exceedingly careful restorations. Here at Summit all the decorations have been refurbished, the irregular glass removed, and the old sashes replaced with hand-molded new ones. Wings have been added, and their interiors have also received this conscientious treatment and boast an extensive array of gougework motifs. Amazingly, the work was done not by professionals, but by the owner, whose modern-day ability would rival that of Springfield's George Chaplin, "the Yankee carpenter who astonished people by his skill."

This bedroom at Summit was added by the owner who did the gougework of the cornice himself. The mantel, window trim, and wainscoting were taken from White Hall, one of the plantations flooded by the Santee-Cooper project. The rice bed is Charleston made (ca. 1790). Beside the bed is a South Carolina-made Chippendale table (ca. 1760) and beneath the window is an English Hepplewhite table (ca. 1800).

Beech Hill

Built ca. 1825, Jono Creek, St. Bartholomew's Parish, Colleton County.

In Colleton County, hard hit by Sherman's forces, Beech Hill alone survives. In 1825 builder Burrell Sanders selected a high ridge running through a tupelo cypress swamp as a site for his two-story black cypress house with broad double piazzas. Unusually wide windows of 12-over-12 panes make the rooms bright and airy. A wide central hall separates the four rooms downstairs, with the same arrangement upstairs. Two interior chimneys conveniently heated the eight rooms. Wainscoting and mantels were not elaborately carved but finished in a neat workmanlike fashion. The second-story hall gives access to the upper piazza through a wide cypress door flanked by side and overhead lights.

A successful planter owning 125 slaves in 1860, Sanders was also a member of the General Assembly and a trustee of the Walterboro Academy. From his Beech Hill plantation account book between 1837 and 1854 we find records of large shipments of rough rice. The rice was apparently raised on Palmetto Island, owned by Wilson Glover or Mrs. Heyward Glover; Sanders leased or supervised the plantation for the Glovers. Later accounts report ginning and packing cotton grown on Beech Hill. The planter's journal gave this highly descriptive weather report 3 April 1854. "It was pritty cold last knigh which produced ice this morning about as thick as a worn 25 cts."

Burrell Sanders died at the age of 80 in 1883. His heirs have farmed the land for four generations and are at present growing fields of crepe myrtles, one of the South's loveliest trees.

Grove

Built ca. 1828, Edisto River, St. Paul's Parish, Charleston County, National Register.

Built about 1828 by George Washington Morris, the Grove is a late example of the Federal style. The son of the Mt. Hope builder, Washington also used octagonal rooms, but here in this raised-cottage design the result is more pleasing. Chamfered columns rest on pillars of brick at the front and on arches at the rear. The molding of the interior is marked by a simple and heavier working of pilasters, wainscoting, and cornice, but above all the change is one of dimension. Here, as at Oak Island, the high ceilings, large-paned windows, and wide, tall doors, are no longer tied to human proportion.

By 1850, Morris had 400 of his 1,112 acres in rice production and owned 136 slaves. After his death the Grove was purchased for $35,000 by John Berkeley Grimball. This Princeton graduate had married Meta, the daughter of Col. Lewis Morris of Willtown. He left in his diary a detailed

The entrance door of this Greek Revival "cottage" is of majestic proportions. The graceful stairway is a modern renovation.

record of plantation life, describing planting, harvesting, and transporting rice, as well as slave care and discipline. He mentions attending the trial of one of his slaves, mediating a duel, and traveling for his health. When the war came, the Grove proved indefensible, the slaves were moved and eventually sold, and the profits were invested in Confederate warbonds—indeed, almost every cent at his family's disposal was converted into bonds. As a result, by the end of the war, Grimball had only a single gold piece given to him by his grandmother with which to buy food. He was, in his own words, "ruined."

A small Confederate battery was located at the Grove during the War. In 1865 the house was shelled, but no serious damage was done.

Several times congregations shifted their services from rural churches to the associated chapels in villages. Here the church building itself was moved, being taken from Willtown in 1879 and reassembled at Adams Run.

Willtown Church and Parsonage

Built ca. 1834, Edisto River, Edisto Island, St. John's Colleton Parish, Charleston County, National Register.

Just to the north of the Mt. Hope house is a single column, all that remains of the Willtown Church; just beyond that is the Willtown Parsonage. An Episcopal Church had been built in 1834 on the site of an earlier Presbyterian building, and the parsonage was constructed that same year. The church, as described at its dedication, was "of singular beauty and completeness," and the minister's residence had "every circumstance considered necessary to his comfort." Unfortunately the congregation dwindled steadily, and in 1879 the church building was moved to Adam's Run.

At the beginning of colonization, Willtown had been second only to Charles Town in size and prospects. The town was divided into 63 blocks, each having four one-acre lots. Provisions were made for an Anglican church, town market, and gardens. At one time 80 dwellings were reported there, but definite records exist for only 11 houses, two churches, and a large store.

Other documents, however, suggest a thriving frontier community. Before 1706, Willtown and Charles Town were the only two precincts in the colony, and several early governors had houses, or at least lots, in both places. Dalcho's first congregation was said to be at Willtown, and the Presbyterian minister Stobo retired there. Although reputed to be difficult to defend in 1715, the town successfully withstood the attack of the Yemassee Indians, and in 1739 the men of the Presbyterian congregation suppressed the Stono slave rebellion.

A free school had been established in 1722, and provisions were also made for roads and mail service. Merchants, carpenters, Indian traders, and even a banker left records of their passing, but despite this promising start the town was floundering by the end of the 1700's. Apparently the lack of a good harbor made competition with Charleston and Beaufort impossible and, like so many other similar enterprises, the town vanished. The church was taken down and reassembled on a smaller scale in the town of Adams Run.

Viewed from a distance is Little Edisto Island and its Windsor house.

Windsor

Built ca. 1857, Russell Creek, St. John's Colleton Parish, Charleston County, National Register.

Facing Russell Creek on what was a part of Little Edisto Plantation, Windsor was built in 1857 and given as a wedding gift from Edward Whaley to his son E. Mikell Whaley. The house is large and simply constructed with high brick piers, following a style that was by then common to the Sea Islands. The spaces between the piers has since been filled in. No doubt the extra height provided by these piers was to protect Windsor from the high tides as well as to catch the breeze.

The interior is divided by wide halls, and the rooms to each side are spacious and have high ceilings. The mantels and wainscoting are simple. Union soldiers occupied the house during the War and carved their names into the hall wall.

Descendant Swinton Whaley was the last to grow Sea Island cotton on Edisto, and he kept a small patch going for seed for some years.

Waccamaw-Pee Dee Rivers System

(GEORGETOWN AREA)

IN 1723 the people of Winyah petitioned to have their inlet declared a port of entry, but it would be more than a century before the full potential of the area was realized. By 1840 the Georgetown district grew almost half of the rice produced in the entire United States, and for that decade and the next more rice was exported from this port than from any other in the world. By Dr. George Roger's estimation, the bulk of the crop was grown by 91 planters, and from 1850 until the Civil War, this aristocracy enjoyed an unprecedented degree of wealth and power.

Almira Coffin, visiting from Maine in 1851, wrote a glowing account of life on the Waccamaw and Peedee.

> We walked from the wharf several hundred yards through a rice field, on a wide bank built up for the purpose of a path, then we came to the high land & entered through a big gate on either side of which was a tree called the "Pride of India" covered with purple flowers & a cherokee rose climbing to the top of one, & a multiflora the other, both having hundreds & perhaps thousands of roses on them. Wild orange hedges as high as my neck & some six feet broad we walked between to the house, which is a two story square one, as large as Aunt B---s, white with green blinds, a Piazza on the north and south sides. The front rooms are used for parlor & drawing room, & any other room, 11 in all, are fitted up for sleeping rooms because they need so many, with all the city company they entertain. There is a large circle in front of each piazza surrounded by hedges, filled with roses, flowering vines & evergreens, each can contain as many flowers as your whole flower garden, and all the trees in the yard are evergreen. Magnolia was in blossom when we left, the first flower of the kind I ever saw, live & water oaks, oranges &c, which makes the place look nearly as inviting in winter as in summer . . . In purchasing a rice plantation they only count the rice land, the high land & buildings are thrown in!

Beneventum

Built ca. 1746, Pee Dee River, Prince George Winyah Parish, Georgetown County.

Built between 1746 and 1756, the original house was apparently only two rooms over two, topped by a hipped roof. At this time the plantation was one of several owned by Christopher Gadsden. This fiery Charleston patriot was a member of the First Continental Congress, and he designed the famous rattlesnake flag which warned, "Don't tread on me." From his hands, the property passed to John Julius Pringle and then, possibly by the marriage of his widow, to the colorful adventurer Joel Roberts Poinsett. Because records are lost, it is

The oldest house in the area, Beneventum means "good or healthful breeze."

impossible to say now who was living in the house when extensive additions were made about 1800.

The rim locks, H and L hinges, and handplaned paneling are all original to the older section of the house. Modern repairs have shown that the plaster laths were split, not sawn, and nailed in place with plantation-made "rosebud" nails. These recent repairs also turned up objects owned by the house's previous occupants. Inside the walls were found a locally manufactured copper dipper, an ivory toothbrush carved with an elephant and inscribed in French, an 1822 newspaper, and a single well-worn wooden shoe.

Arcadia

Built ca. 1794, Waccamaw River, All Saints Waccamaw Parish, Georgetown County, National Register.

In 1769, Joseph Allston acquired a tract of land from Anthony and George Pawley, and named the southern portion Prospect Hill. His son Thomas inherited the property and apparently had begun construction of the house by 1794, for in that year he bequeathed to his widow, Mary, the plantation and "house frame."

The house, which Mary completed, is similar in some respects to the Miles Brewton house in Charleston. The double-tiered portico with its unusual Ionic columns and the hipped roof with modillion trim are particularly suggestive of this earlier building which was owned at the time by Joseph's brother, Col. William Alston (he dropped the second l from his name in 1791). William is said to have built an identical house at neighboring Clifton, but that house burned almost immediately after its completion.

The Prospect Hill house, however, is certainly distinctive in its own right. The fine double entry steps are the earliest of this design we have encountered, and the exceptional interior of the house has a thoroughly "Federal" complement of dentil cornice with fretwork, paneled wainscoting, and frieze-embellished mantels in the Adam style.

Mary Alston sold the property to Joshua John Ward. Although her late husband, Joseph, had been one of the area's most successful planters, owning 203 slaves in 1790, it was under the direction of the Ward family that peak rice production

An observer remembered Arcadia in the 1820's: "This was a show place, conspicuous in the southern states; the mansion house was large and handsomely furnished."

The same visitor to Arcadia remarked: "and the adjacent grounds in which were many rare plants were kept in beautiful order."

was reached at Prospect Hill. In 1850 their large holdings were worked by 1,100 slaves and yielded 3,900,000 pounds of rice. In 1860, they produced the most rice ever grown in the district, 4,410,000 pounds.

In 1906 Dr. Isaac E. Emerson bought Prospect Hill, gradually acquired several adjoining plantations, and renamed the whole property Arcadia. Several unusual but fitting additions have been made to the building during this century, and the terraced grounds have been refurbished.

Litchfield

Built ca. 1794, Waccamaw River, All Saints Waccamaw Parish, Georgetown County.

A 1794 plat shows a sketch of an avenue and a house with chimneys at each end. It is possible that the builder of Litchfield was Peter Simons, for his son John inherited the plantation at that time and sold it to the successful Georgetown merchant, Daniel Tucker. Tucker too has been suggested as the builder, but he died the year after he bought the property.

Tucker's eldest son, John Hyrne Tucker, acquired the plantation and added much to the building and grounds during his long tenure there. A graduate of Brown University in Rhode Island, he was quite successful as a planter. F. A. Porcher recalls:

> Rice planting was his sole delight. He lived for and in rice. It was the first and the last thought of his mind. He had exquisite taste in wines. He had a large stock of the best Madeira, and was almost as proud of his wine as of his crops. He was superstitiously religious regarding the Episcopal Church as the only true and safe road to Heaven.

Obviously rice planting wasn't quite his sole delight, for besides wine and the Church, Porcher also mentions his "large and promising family of grandchildren." Although pitted by smallpox and having "an enormous nose full of blue veins and a knob on the end of it," Tucker married four times and fathered nine children.

Litchfield's unusual central room is well-lit and handsomely furnished.

Now used as a guest house for the Litchfield Plantation Community, this was once the home of John Hyrne Tucker, "a good man, and an honest man, and a happy man."

A son by John Tucker's third wife Susan Harriet Ramsay, Henry Massingberd Tucker was the next owner. Educated as a physician, he was also an outstanding planter. His father had grown 1,140,000 pounds of rice in 1850, but in 1860 the son grew 1,500,000 pounds. Remembered also as an excellent sportsman, Henry was such an unbeatable marksman that he gave up competing for the prizes of the Georgetown Rifle Club.

From him the plantation passed through the Lachicottes and Wards to Dr. Henry Norris, who restored the building.

Chicora Wood

Built ca. 1819, Pee Dee River, Prince George Winyah Parish, Georgetown County, National Register.

Until recently Charlotte Ann Allston was credited with building in about 1819 the main part of the present Chicora Wood house. However, this would have been unlikely since her letters to her son during that period sound thoroughly beleaguered. "Battling with the wide, wide, unfriendly world," she wrote. "People here, whose chief object is to make Rice to buy Negroes and Buy Negroes to make Rice," she observed at another point. "Think of it my Dear Robert, your Father split upon this very Rock."

The current owners, using an early plat, deduced that the rear wing of the house was built by Charlotte's husband Benjamin Allston before his death in 1809, and that their son Robert Francis Withers Allston built the major front section in the late 1830's. Robert's daughter Elizabeth Allston Pringle confirms this:

> Papa found the house at Chicora too small for the growing family, and began the planning of a new one, to which the two very large downstairs rooms of the old one should be attached as an L.

Architect Randolph Martz has been able to reconstruct this original "old" building and found it to be two 20 foot square rooms on either side of a central hall with two chimneys on the land side. This was rearranged by Allston's renovation.

The front of the main house, a two-story, gable-roofed building on a raised brick basement, is surrounded on three sides by a piazza that has been called "the finest in the Waccamaw region." The interior, though badly vandalized and stripped of its mahogany after the War, still retains some of its original trim—most notably, cast-plaster cornices and marbleized baseboards on the second floor. The two rooms to the right of the entrance hall are joined by two double sets of great doors to approximate the twin parlor arrangement popular for entertaining in its day.

Robert Francis Withers Allston was born in 1801, the son of Benjamin Allston, Jr., and Charlotte Ann Allston, who were cousins. After his father's death in 1809, his mother took over the management of the Pee Dee rice plantation Matanzas, which we know today as Chicora Wood. Robert was educated in Georgetown, and at 16 entered West Point. Upon graduation he took a military commission, but resigned it in 1822, returning home to assist his mother.

Charleston belle Adele Petigru accepted the proposal of the "obstinate" Allston in 1832, and the couple had ten children, five of whom reached adulthood. After struggling for some 14 years, Allston began in 1840 to acquire more land and eventually owned seven plantations, two houses in Georgetown, and the Nathaniel Russell house in Charleston. Active in social and civic circles, he served for the States Rights Party in the S.C.

***Built by Governor R. F. W. Allston, handsome Chicora Wood was celebrated by his daughter Elizabeth in* Chronicles of Chicora Wood.**

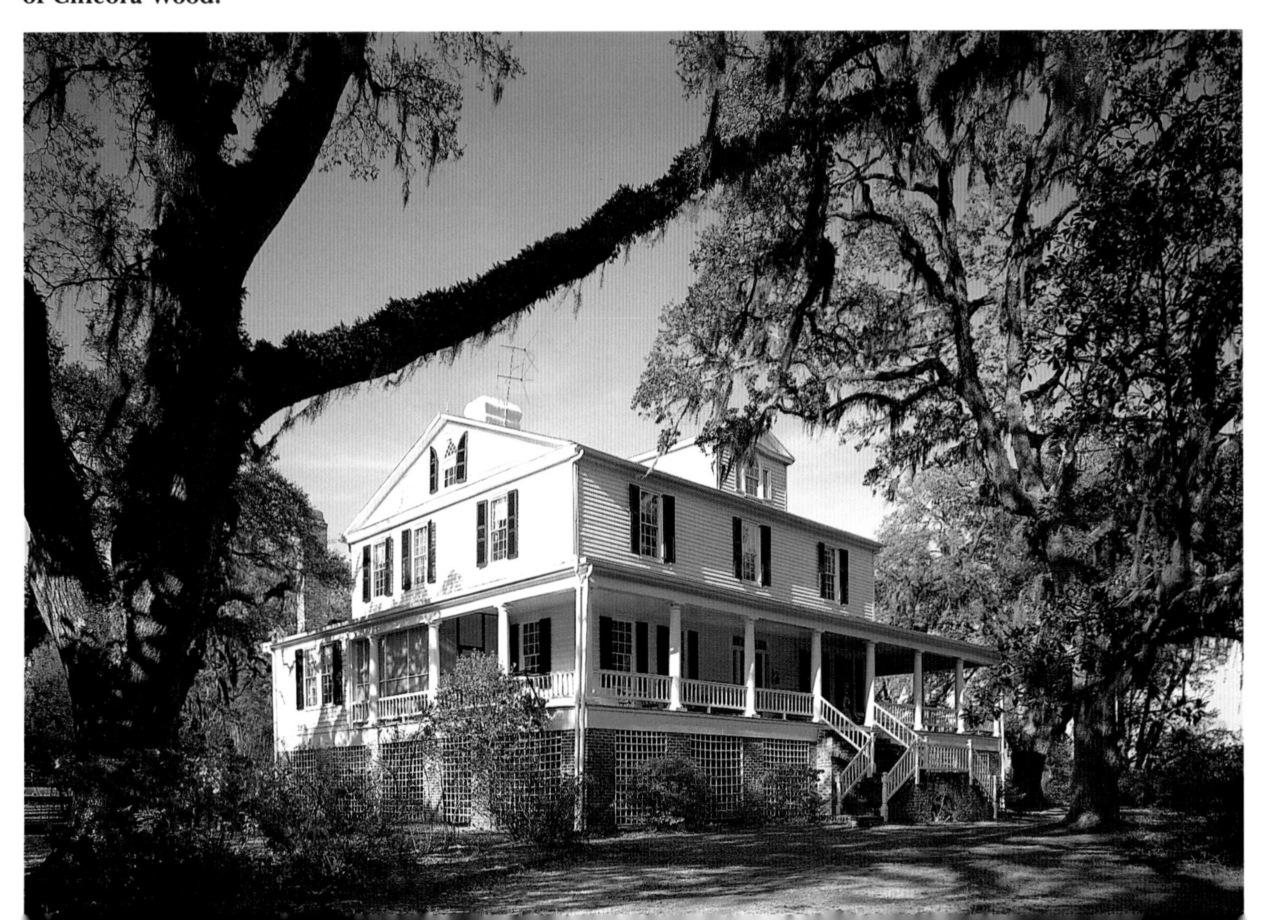

"We looked at the house; it was a wreck—the front steps gone, not a door or shutter left, and not a sash. They had torn out all the mahogany panels below the windows and above the doors there were panels painted—the mahogany banisters to the staircase going upstairs; everything that could be torn away was gone." Much of the damage has been repaired since daughter Elizabeth's 1865 description. The width of the hall suggests the spacious dimensions of the whole. The floor boards run the length unbroken.

House and Senate, and in 1856 was elected Governor on a progressive platform advocating better public education. Too old to fight in the Civil War, he ran the plantations of his sons and nephews, but died in 1864 knowing the cause he had advocated so strongly was lost.

An intimate portrait of life on the plantation was left by its next resident, Allston's daughter Elizabeth Allston Pringle. In *Chronicles of Chicora Wood* she tells of domestic life, summers at nearby Pawley's Island, the social life of Charleston, and the tragic deaths of her siblings. Surviving the hardships of the War, she and her mother singlehandedly reclaimed their properties from their former slaves, only to have all but Chicora Wood confiscated for a family debt Allston had generously assumed in the previous decade. Despite the setback, Elizabeth eventually succeeded at the occupation her grandmother had pleaded so fervently against, that of rice planter.

Exchange

Built ca. 1825, Pee Dee River, Prince George Winyah Parish, Georgetown County.

Probably one of the oldest houses along the Pee Dee, Exchange appears in the Mills Atlas of 1825 as belonging to Davidson McDowell. Family tradition holds that it gained its name because it was exchanged for another piece of property. If so, the property exchanged may have been part of Mc-

A Victorian mantel frames the small fireplace, appropriately scaled for this comfortable room.

Dowell's purchase of neighboring Rose Bank in 1819.

In 1822 the owner married Mary Moore of Charleston, but she died the following year on the Waccamaw seashore. In 1827 he married the widow of a Williamsburg planter, Catherine McCrea Witherspoon, with whom he had six children. Ten years later he sold the plantation to John Weston, whose heirs transferred it in 1843 to Robert F. W. Allston of neighboring Chicora Wood. Sold and rebought, it was part of authoress Elizabeth Allston Pringle's canceled legacy. When the estate was settled, her two brothers, Benjamin and Charles, purchased it; in 1899, like Arundel, Exchange was taken over by the Guendolas Rice Company and passed on to the LaBruce Family.

Of uncertain date, this house might have been extensively renovated by Allston's large crew of slave carpenters. The Doric columns and front dormers suggest at least some link to Chicora Wood.

The mantel and window framing of Exchange's dining room appear original to the building.

Dating from the ownership of Dr. Parker, six small slave cabins and the church school on the right make up "the Mansfield Street."

Mansfield Winnowing House and Slave Row

Built ca. 1835, Black River, Prince George Winyah Parish, Georgetown County, National Register.

In 1756, widow Susannah Man purchased the plantation that would remain in her family until 1912. It passed from her to her grandson Archibald Taylor and then to his son John Man, and eventually to John's daughter Anna. Anna's daughter, Mary Lance, married Dr. Francis S. Parker in 1835, and he traded Wedgefield to his father-in-law for Mansfield. It was while this last couple owned Mansfield that rice production reached its peak, and it is with their names that the property is most often associated.

Dr. Parker gave up practicing medicine and de-

Mansfield's winnowing house is the only one still standing in Georgetown County. After the rice was threshed it was dropped through the grating at the top of the structure, and the wind separated chaff from seed. This operation was particularly important for the gathering of unbruised seed rice. In the background is the stack of the recently burned threshing mill. Run by steam, the threshing operations removed the head from the rice stalk.

voted himself entirely to planting. He kept plantation records during the latter part of his tenure that indicate that not only did he plant 219 acres here, but that he had two other plantations as well; in 1860 he grew almost one-and-a-half million pounds of rice.

Dr. Parker voted for the Ordinance of Secession, and during the War served as provost marshal of Georgetown. Although ruthless by modern standards, Parker's contemporaries thought that he performed this last job "with great decision and judgement." His "palatial home" was reportedly destroyed at the end of the War, so the existing house was presumably built soon afterwards. The slave street and winnowing house, however, are original. The rice mill, restored to full operation in 1943, recently burned.

Arundel

Built ca. 1841, Pee Dee River, Prince George Winyah Parish, Georgetown County.

As with neighboring Dirleton and nearby Annandale, Arundel owes its name to the surge of Romanticism that swept over the South in the early 1800's. Inspired by the novels of Sir Walter Scott, the planters put aside their Indian and Carolinian names and went back to the castles of an almost mythical Scotland and England. Here the source is

Framed by the heavy pilaster of this Victorian doorway is a portrait of one of the present owner's ancestors. Ceiling medallion, bay, and unusual eight-over-eight-light-sash windows are original to the Arundel house.

Arundel Castle on the Arundel River in Sussex, England.

When bought by Dr. William Allston from John Julius Pringle in 1806, the property contained a house, which burned in 1827. In 1841 a northerner, Frederick Shaffer, purchased the property and began building the present dwelling. He never lived there, however, and in 1860 it was sold to Charles Allston and then to Louis Lachicotte, who deeded it to the Guendolas Rice Company, which was owned by himself, his brothers, and Louis LaBruce. This company, like several others, hoped to cut costs by accumulating large acreage (in this case six plantations) and by doing its own milling. LaBruce eventually owned all of the property and it was he who finally finished the house at the end of the century. After standing vacant for 60 years, Arundel was finally occupied.

The present house is believed to have been begun about 1841, although it was not completed until the 1890's. The heavy, deeply molded trim associated with the high Victorian period has been beautifully refinished. The original one-story front porch has been replaced by a portico with monumental columns.

An assortment of original buildings on the grounds includes a Gothic smokehouse, commissary, watchman's camp, and an overseer's house dating from the original building of 1790. An extensive garden is shaded by great oaks, and the main entrance is guarded by gateposts of original design, appropriately decorated with acorns.

Dirleton

Built ca. 1856, Pee Dee River, Prince George Winyah Parish, Georgetown County, Owned by the State of South Carolina.

Dr. James R. Sparkman, who had an interest in architecture, designed Dirleton to suit his country lifestyle. Low to the ground, the house in some ways appears to be a Gothic Revival version of nearby Chicora Wood. The piazza encircles the house, but here has square columns. Wide eaves overhang the gable ends and project from the front of the house at a barely noticeable slant. Only one front dormer appears in an early photo, so all may have been later additions. The same photo shows the house with white trim, shrubbery, and occupants—altogether a less austere Dirleton than we see today.

The six-over-six high windows are oversized, and from doors to baseboard the simple interior is on a grand scale. As with Chicora Wood there is a twin parlor to the right of the central hall. The two rooms join here without the interruption of a fireplace. The fireplaces, with their individualized marble mantels, are to the side on four separate chimneys. The flooring is narrow and the high ceilings are bare except for a single plaster circle decorating each room.

The house shows the influence of the late nineteenth century; although started in 1856, it was not completed until after the War. The builder acquired the plantation on his marriage to Mary Elizabeth Heriot, whose family had named it after their Scottish home.

Dr. Sparkman was a gifted man of broad interests. His medical practice before the War included the care of over 3,000 slaves, as well as masters and their families, yet he still had time for extensive rice planting on several properties; in 1860 he grew almost a million pounds of rice. After the War he was one of the first to try large-scale rice production with contracted labor. These efforts were unsuccessful, but he continued a long and relatively prosperous medical practice.

In 1858 he wrote a report for the state based on his 100 slaves at Dirleton, attempting to describe slavery on the average rice plantation, including a detailed account of work, food, housing, medical and spiritual needs. After the War he took an active interest in restoring the South, but beyond the paternalistic slave system pictured in his report he could see no solution to the race problem. In 1889, shortly before his death, he wrote a final pessimistic summation that bemoaned the prevailing state of affairs, ending with a plea for the speedy education of "a semi-heathen race."

A daylight apparition, only Prince Frederick's belltower remains. Built on the eve of the War, the building was the only parish church built in the Gothic style.

Prince Frederick's Church Ruins

Built ca. 1859, Pee Dee River, Prince George Winyah Parish, Georgetown County.

Originally a part of Prince George Parish, Prince Frederick's was established as a separate parish church in 1734. At first the congregation met at an existing building on the Black River, but in 1835 Rev. Hugh Fraser donated the present site near the Pee Dee. Fraser was an outstanding priest and enjoyed a success denied to most others. The building of a new church in 1837 coincided with the rising prosperity of the rice planters.

Prince Frederick's received a new rector in 1846. This man, Joseph Hunter, was so popular that by 1859 it became apparent that a new church was necessary to accommodate the increasing number of communicants.

The cornerstone of the building was laid in 1859, with Governor Robert F. W. Allston delivering the address. The Union blockade, however, slowed construction. When Gunn, the contractor, was killed (the church is also known as Gunn Church) work stopped altogether. The unfinished church was damaged during the War, and it was not until 1876 that a generous gift made completion possible.

By the 1950's it seemed that the building was beyond repair and the grounds were cleared and fenced. Church officials, fearing the collapsing building would hurt someone, had all but the belltower demolished.

Black Mingo

Early in the 1700's settlers pushed inland, following the Black and Pee Dee Rivers, while others of Welsh and Scotch Irish extraction traveled down the seaboard. Far removed from markets and the tidal flow necessary for large-scale rice production, the upriver planters successfully raised hogs, corn, and indigo. Finally in the 1800's cotton brought real wealth to the area.

Even the narrow Black Mingo was traversed by steamboat, thus connecting inland planters to the coast. A major stage line also passed through the area, but no travelers came this way without complaint. As late as 1816, overland travel in the Low Country was uncomfortable and frequently hazardous. One northern traveler described his travels this way:

> I was dozing in the dark when I was awakened by the voice of the driver, vowing that nothing should tempt him to encounter a danger like that he had just escaped. He had passed one bridge, another remained, and he kept his vow: but what was to be done to escape sleeping in the woods? The bridge might be avoided by an old road through a swamp, supposed to be impassable: here, however, we were to make the attempt. Branches of pine were cut and lighted for torches, and we proceeded through the woods. After some mistakes and more oathes we found the bog, which indicated we were in the right way—"to be upset," I said to myself; but we dashed through it up to the traces, with crash, whip, and halloo. Such an equipage, in such a place, with torches, and negroes, and harsh sounds, more resembled a vehicle for the transport of the damned to their infernal dwelling, than a stagecoach in a rational country.

William Cooper House

Built ca. 1758, Prince Frederick's Parish, Williamsburg County.

To appreciate the William Cooper House we should begin with a 1733 description of the neighborhood:

> Expectant that we were coming to a very agreeable place, but when we arrived and saw nothing but a wilderness and instead of a fine timbered house nothing but a very mean dirt house.

By 1758, William Cooper had built a house for his wife Jane James that would indicate a certain degree of wealth and civilization had reached Black Mingo.

The original building was a simple two rooms over two with a piazza across the front. Both inside and out, the work was done with careful attention to detail and an ingenious use of available material. The scored stucco finish of the chimney, small candlelight dentil running beneath the eaves, beaded flushboard beneath the porch, and an unu-

Still in place, the corner cupboards of William Cooper's house have their original glass. Such care was used in the marbleizing of the woodwork, that even after two centuries of wear it must be touched to be disbelieved.

sual diamond-inscribed front door were exceptional in what was then a frontier.

The interior was more remarkable still. Inside there is no evidence of plaster: the building was finished completely with cypress boards, most 23 inches wide. For wainscoting, these boards were run horizontally and scribed in panel-like sections. Apparently only the wainscoting was gilded and marbleized, and the remaining paneling was run vertically above, and perhaps left unpainted. The plank ceiling was trimmed with a wide, elaborate cornice. Today the corner cupboards still have their original glass, and the mantels reveal fine workmanship, with the overmantel of the main room made from a single piece of cypress.

At one time a hall was added to the center of the house, but the entry was later returned to the larger of the two rooms. The older stairwell towards the front of the smaller room gave access to rooms that were decorated as painstakingly as those below.

William Cooper left the house to his son George, who married Thermuthis Ann Montgomery. Between 1847 and 1854 George had his slaves build four houses for his children, and all four dwellings have survived, three still inhabited by his descendants. The William Cooper House is occupied by the builder's great-great-great-great-grandson.

Rodger McGill House

Built ca. 1767, Prince Frederick's Parish, Williamsburg County.

Rodger McGill married Elizabeth Westbury on February 23, 1767 and left in the family Bible a record of his house's first residents. Spaced two years apart, nine childen were born to the couple before Elizabeth died in childbirth with twins. The Bible reads:

> No monument her virtues can supply
> In the cold grave her fair body lies.

If not a monument, at least a fine reminder of Elizabeth's virtues is this sturdy home that today still belongs to her descendants.

In 1825 the house was moved three-quarters of a mile south of its Indian Town Swamp site. Apparently the structure remained unaltered—the small windows of the second story, for example, indicate that a piazza has always encircled the dwelling. The columns are of particular interest, for they were hand-hewn rather than turned on a lathe. The accompanying banister was copied from a small original section that survived.

In this early frontier home hand-hewn columns stand free of an encircling piazza. Swamp land was diked for growing rice, but as on most of the neighboring properties, it was cotton that brought prosperity to the McGill family.

As in other houses in the area, the columns of the Rodger McGill house stand free of the balustrade, and above the upstairs fireplaces the chimneys separate from the edge of the house, clearing the eave at the peak of the roof. (Seen also at the Cooper House and the Motte residence in Mt. Pleasant, this construction was apparently intended to avoid having to flash the roof.)

The interior, consisting of three rooms over two, was finished completely in cypress, as was the Coopers' house. But here the work was not so polished, and at times the framing was left revealed.

A grandson of the builder, Dr. Samuel McGill wrote *Reminiscence of Williamsburg County*, a thorough and entertaining account of the area, which surely deserves a wider audience. He speaks of the "joyous excitement" the Cooper and McGill children felt when the two families met, for his parents:

> loved the Coopers' at that time, and having known them in their infancy they were even on intimate terms. Col. William Cooper and Mr. George Cooper were much concerned in his father's behalf, when he was summoned to trial in the church session house for dancing and giving dancing parties in his house.

China Grove

Built ca. 1787, Prince Frederick's Parish, Williamsburg County.

In South Carolina the Revolution was won mainly in the back country, and the active involvement in the war of these frontier settlers forced the more established planters of the coast to share political power with them after independence. One skirmish was fought at Rodger McGill's house, and another at the site of the Black Mingo Church. The builder of China Grove, James Snow, is on record as having supplied Francis Marion's Brigade and Peter Horry "with cattle, many pounds of beef, horses, and clean rice, many bushels, 1780, 1781, 1782." Nearby Snow's Island was often used by the Swamp Fox as a refuge during his guerrilla campaign, and it also became a scene of battle.

Snow received a grant for his 1,000 acres in 1787; if not built then, the house was finished soon after the Revolution. The simple farmhouse design is common in the area. Four rooms below are divided by a hall, where a straight stairway leads to two rooms above. A shed over the rear rooms is balanced in front by a porch. Here we find free-standing columns, and beaded flushboard siding beneath them.

The interior walls are plastered and trimmed simply with 12-inch beaded baseboard and plain cornices. Paneling beneath the windows is original to some rooms. The doors also date from construction, and until their restoration in the 1940's had string latches.

Since China Grove is unusually far from a river, it seems doubtful that rice could have been grown on the plantation in large quantities, but indigo was. Cotton, farmed well into this century, was the crop of descendant James Snow, III. He gained the disapproval of his neighbors by using 77 freed men instead of slaves to work his land. The property remained in the Snow family until 1897, but fared poorly in the years after. In 1948, the current owners rescued the abandoned building and turned it into a comfortable home.

Black Mingo Church

Built ca. 1843, Prince Frederick's Parish, Williamsburg County, National Register.

Built in 1843, Black Mingo Church is a good example of that often cited but seldom seen indigenous genius—American architecture. The low-pitched roof, pediment, deep entablature, and pilasters at the corners outline the meeting house in the classic mode of the Greek Revival. The siding of narrow cypress is broken by small-paned round-headed windows of the Federal period. Neatly lettered Scripture across the front and side frieze complete what—taken in this wilderness setting—can only be described as a blessed incongruity.

Inside, two lines of straight-backed pews and a U-shaped balcony crowd beneath a narrow coved ceiling, focusing attention on a massive pulpit whose dark stone-like bulk is emphasized by the back light of a single great window. Here, too, are

A small Baptist meeting house is disguised as a temple, or vice versa. Above the door of Belin's Black Mingo Church a precisely lettered inscription invites, "Come unto me all ye that labour and are heavy laden and I will give you rest."

inscriptions admonishing worshippers to follow the ways of the Lord. Printed inside the pulpit where only the preacher can see is a reference to Ezekiel, warning him to take stock of himself before dealing with the wicked.

Cleland Belin, leading merchant and Baptist layman, financed and supervised construction of the building that sometimes bears his name. After running away from his Georgetown home, he settled as a young man in Willtown on Black Mingo Creek. The community had thrived for a time as a "place of some business" but a political squabble caused the public road, a chief artery of traffic on the eastern seaboard, to be cut off. Though the town was dying, Belin flourished as a merchant and built for himself a mansion that was the marvel of the locale. Each of its 12 rooms was said to contain a grandfather's clock set at a slightly different time, so that the inhabitants were continually being entertained by chimes and "dancing fairies."

Famed for his hospitality, Belin entertained lavishly. The governor was a frequent visitor, and horse racing a favorite amusement. Although any advance in the technology of his age brought Belin joy, it was apparently the church that gave him the greatest pleasure, and he is said to have stopped there each morning before beginning the day's activities.

Belin and his wife Sarah Margaret McFadden had 13 children, 11 of whom died before the age of five, presumably the victims of diphtheria, malaria, and typhus epidemics. Belin himself died in 1868, and is buried along with much of his family on the church grounds. All other evidence of Willtown is gone.

The choice timbers used in building the church came from the mill of John Gordon, Jr., and although not documented, legend has it that Gordon's sons James and David professionally executed the lettering of the Scriptural verses.

Reached by a small iron stair, the pulpit was said to have at one time been higher. Each pew back is carved from a single piece of lumber with the curve of the log serving as the rounded top.

Port Royal Sound

SPANIARDS, Frenchmen, Indians, British, and Union troops have at one time or another occupied this area, and each in their own way laid waste to the countryside. As a result of these depredations, only a handful of plantation houses have survived, the majority of these on St. Helena Island. Here, as on Edisto, indigo was grown with success, but it was Sea Island cotton that brought great wealth. Beaufort, officially established in 1711, could not compete with Charleston as a port, but suited the local planters well as a summer retreat and dozens of their fine townhouses still stand.

While exploring the area in 1665, William Hilton made the following report:

> The country abounds with Grapes, large Figs, and Peaches; the Woods with Deer, Conies, Turkeys, Quails, Curlues, Plovers, Teile, Herons; and as the Indians say, in Winter, with Swans, Geese, Cranes, Duck and Mallard, and innumerable of other water-fowls, whose names we know not, which lie in the Rivers, Marshes, and on the Sands: Oysters in abundance, with great store of Muscles; A sort of fair Crabs, and a round Shel-fish called Horse-feet; the Rivers stored plentifully with Fish that we saw play and leap.

Two centuries later Rev. J. J. Sams repeated this appraisal,

> Datha was to us a kind of terrestrial paradise. With the help of sword and cannon and foreign soldiery, the Yankee people have wrested it from us. They have impoverished a rich [land], rendered unhappy a happy family, scattered a united family and deprived you of your inheritence, small as it was. You must forgive them and I forgive them. I do forgive them, for I know I need forgiveness.

Retreat

Built ca. 1745, Battery Creek, St. Helena's Parish, Beaufort County.

When the mysterious French merchant Jean de la Gaye sailed for England in 1769, he left behind a clouded estate that included a small story-and-a-half house, thought to date from the early 1700's. Besides being the oldest structure south of Charleston, it is also one of the few tabby homes to remain intact. Twenty-two inches thick, the scored stuccoed walls of oyster shell and lime are interlocked at each end with the fine brickwork of the chimneys. Small "English" brick is laid only to the first story level, and a different decorative diaper has been set with blue "glazers." Inside, the two smaller rooms have corner fireplaces.

Jean de la Gaye, a member of the nearby Purrysburg colony is credited with building the house, but if this is so it would have been after 1748 when he purchased the tract on Palmentor's Creek. It seems more likely that the house was constructed earlier by Thomas Simons. Simons purchased 250 acres from William Ford in 1712, paying £55. His son, Thomas, apparently sold the property to de la Gaye in 1745 for £500.

There is a romantic tale told of the de la Gayes. After his wife's tragic death, the heirless Frenchman lived on in embittered seclusion until murdered by two of his slaves (one of whose head lends its name to nearby Skull Creek). Correspondence with Henry Laurens, however, shows that de la

Gaye returned to France with his wife and remained there. In contrast to the misery of the legend, the letters concern themselves with the purchase of handkerchiefs and the lawsuits that de la Gaye had left behind.

Following the Civil War, the house was bought for $8 in unpaid taxes. Its condition gradually deteriorated until, by 1939, the roof had fallen in and the wainscoting and flooring were being ripped out. "Save the Retreat before it is too late," a concerned reporter wrote at that time, and it is our good fortune that someone did.

This small cater-cornered fireplace is typical of early colonial construction. Horses, seen through the window, graze where merchant de la Gaye once tended his vineyard.

Retreat's tabby walls and chimneys are obviously the work of skilled hands. The mason left his mark outlined in glazed blue brick across each end of the house.

Built on ground hard won from the Yemassee, and then burned by both British and Union armies, Prince William's was one of the most impressive of the rural churches. The engaged columns running its length make it the earliest building in America to emulate a Greek temple.

Prince William's Church (Sheldon)

Built ca. 1751, Prince William's Parish, Beaufort County, National Register.

Here is a vivid reminder that the Low Country was often a battleground. The Yemassee had decimated this area in 1715, but by 1736 we find a reference to a chapel "on the South Side of Combahee River, near Hoospa Neck." Nine years later the Parish of Prince William's was split from St. Helena, and William Bull, Stephen Bull, Robert Thorpe, James Deveaux, and John Greene were named as commissioners to build a church and parsonage.

Gov. William Bull, active in both Proprietary and Royal governments, was the guiding hand behind the project. Once the church was completed, his son Stephen entertained on the Sabbath. "At his expense," the entire congregation participated, arriving in at least 60 or 70 carriages. The local nobility won out over Prince William, Duke of Cumberland, and the church was called Sheldon after the adjacent Bull estate.

During the Revolution, ammunition and arms were hidden in the Bull family vault. In 1779 General Prevost's army, marching north from Florida, burned the church. It remained in ruins until 1824, when it was rebuilt along its original lines. Sam Stoney quotes William John Grayson who claimed they had turned a "beautiful ruin into a very ugly church." The building was returned to its ruinous state by Union troops, who used the building for a stable and then burned it.

What remains of Sheldon is certainly awe-inspiring, not only for its size, but for its grace and proportion as well. The columns of the vanished portico are of carefully molded brick, and six engaged columns run down each side making this one of the earliest uses of the temple form in America.

Mr. Stoney in his investigation pointed out the "put-log holes." In early construction, one end of the scaffolding timbers was commonly placed inside the wall, and the resulting hole plugged when the building was finished. Two fires have caused these to reappear, Stoney notes. He then relates a second discovery: "I was eating my lunch, not wasting time, you see, and sitting on a dead Bull, it suddenly came to me that the pattern of glazed headers was the design of the year of construction, 1751." Just off the highway at Garden's Corner, the church is easily found, and it is worth a trip to see this imaginative and no doubt correct interpretation of the staggered glazers.

A little easier to spot are the blue checkerboard bonding bricks on the far side. Fire and weather have apparently worn away the rest, but at least this example of uncommon decoration remains. "The second best church in the province, and by many esteemed a more beautiful building than St. Philip's," reads the Anglican report to England, "beautifully pew'd and ornamented."

Dataw

Built ca. 1783, Colleton River, St. Helena's Parish, Beaufort County

An Indian captive taken by the Spanish told of Dataw, a giant Indian King, whose size was due to his eating special herbs. The island, sometimes also known as Datha, was named in his honor. It became an early Indian trading post and in 1698 was granted. Used in the succeeding years to graze cattle and grow indigo, it was later sold by Lewis Reeves to his cousin William Sams and his wife Elizabeth, originally from Wadmalaw. William had been sympathetic to the British and moved in search of "a more comfortable climate." The island was eventually split between two of their sons.

Dr. Berners Barnwell Sams received this portion and the small tabby house that may have dated from before the Reeves occupancy. B. B., as he was called (his numerous children were also named with double initials), added an identical wing off the back corner of each side, and a passageway across the rear which connected all three buildings.

Various outbuildings were built in the surrounding area, the most substantial being of tabby. Among these were an overseer's house, slave cabins, kitchen with eight-foot fireplace, smokehouse with unique tabby roof, dairy, fowl house, and large pigeon house.

Sams was a respected doctor, tending carefully to his own slaves and to others from his Beaufort townhouse, but he was most noted as a planter. He planted indigo and cotton with great success, but most unusual were his orchards. Besides 35 acres of oranges, he raised pears, figs, apples, and plums. His son, Dr. J. J. Sams, wrote an interesting memoir of his boyhood on the island and left of his father a description of an able administrator and a man of great practical knowledge. "He owned tailors, blacksmiths, and carpenters, but he seemed always to know more about these trades than the servants themselves."

Tabby construction was an art to which Sams devoted particular attention. This concrete-like mixture was used on the Sea Islands where clay for brickmaking was scarce and oysters plentiful. The recipe was one part shell-lime, two parts sand, and four to six parts oyster shell. The doctor's son explains that this mixture, moistened with water, was packed into boxes 15′ to 20′ long and allowed to dry. "The sides and ends of the boxes were held by pins. . . . The box was taken down and put upon the tabby already dry, and so box after box was packed or pestled until the walls were as high as you designed."

Nowhere was this art practiced better, but unfortunately the house accidentally burned in 1876.

Tombee

Built ca. 1795, Station Creek, St. Helena Island, St. Helena's Parish, Beaufort County, National Register.

The oldest existing house on St. Helena Island, Tombee takes its name from its builder Thomas B. Chaplin, Sr. Long-staple cotton was first planted on the Sea Islands in 1790, and since a small foot-pedal gin was already in use before Whitney's, profits were realized early.

Chaplin may have built this house as early as 1795. The base of the tabby foundation is a great slab that lies four feet below grade. Placed close by Station Creek, the house's unusual T-shaped floor plan gives each room exposure on three sides, and of course, the double-tiered wide piazzas face into the prevailing sea breeze.

John and Phoebe Chaplin came to the island early and were there when the Yemassee War began. Probably raising cattle and later indigo, they must have been successful, for their son, Tom B., the builder of Tombee, is listed in 1790 as already owning 65 slaves. Living in this same house, his grandson Thomas B. Chaplin did not fare so well. The comprehensive journal he kept between 1845 and 1858 makes clear that simply owning a plantation did not guarantee prosperity. He included in this volume notes on the weather and crops and a detailed description of social life on St. Helena.

His own struggle began when he married at age

An unusual T-shaped floor plan gives good ventilation to all rooms of Tombee.

The 1840 mahogany table and silver service are family heirlooms of the present owner.

16 and was installed at Tombee by his mother. Through inattention and preference for brandy, cards, and billiards, he neglected the plantation, while his neighbors planted every available square foot and became rich. To worsen matters, his mother married a man 20 years her junior, and a long-running lawsuit against his stepfather began. Unable to manage his slaves efficiently, he began to sell them to pay his creditors, and at perhaps his lowest point he lost his summer house at St. Helenaville and looked on in desperate helplessness as his children fell victim to fever.

Not all was bleak, however, for he hunted and enjoyed long fishing trips with his family, and caroused with fellow members of the militia and St. Helena Agricultural Society. (He was sore pressed when it was his turn to "find" dinner for the latter.) At the death of his invalid wife he married his sister-in-law, long-time resident of the house. At last the tide of ill-fortune turned and he was beginning to succeed, when the invasion of Port Royal during the War between the States brought an abrupt end to his life as a planter.

In his journal Chaplin complained about the house he had inherited. He could find no privacy in the six main rooms, for it seemed his five children, wife, sister-in-law, and numerous servants were always underfoot. Less legitimate is his claim of "a miserably constructed roof"; references to broken windowpanes and decrepit stairs indicate that he was not swift to make normal repairs. Already slipping into disrepair while Chaplin lived there, the house was eventually abandoned. Early photographs show that the house's present immaculate condition is the result of a masterful restoration.

The finely done but relatively subdued use of gougework suggests an early date for this Federal interior. The wainscoting and mantels have been given a careful feather-grained finish. Topped by a glass cabinet, the desk is eighteenth-century American made. The rug is an early-nineteenth-century Heriz.

Located at the northern end of St. Helena Island, Coffin Point could well have been where Spaniard Francisco Gardillo stepped ashore on St. Helen's Day, August 18, 1520.

Coffin Point

Built ca. 1800, St. Helena Sound, St. Helena Island, St. Helena's Parish, Beaufort County, National Register.

Located at the opposite end from Tombee on St. Helena Island, this more ambitious house looks out on Port Royal Sound and the Atlantic Ocean. The foundation here is tabby and stands taller than at Tombee. In place of the upper porch is a deck. Trimmed with dentil cornice, the roof is hipped with a central pediment and with dormers off each end. As opposed to Tombee's single rectangular transom with inserted fanlight, the main door of Coffin Point is itself semi-elliptical, and the door to the upper deck has side and fanlights typical of the Federal period. The interior, finished with wainscoting and dentil-trimmed mantels, is distinguished by its embellished archways and its scroll-trimmed stairway that rises to the dormer rooms.

The builder, Ebenezer Coffin, kept extensive plantation journals. An 1800 entry notes that he had engaged Mr. Wade and five other carpenters "to work in St. Helena in erecting a dwelling house, stable, negro houses." In sharp contrast to the hapless Chaplin, the 28-year-old Bostonian was well on his way to having "the best managed and most prosperous plantation on St. Helena."

A friend of the successful William Elliott of Hilton Head, Coffin too was a scientific agriculturlist whose cotton brought top prices and whose seed was prized by his neighbors. Judging by his account books, he ran the plantation with a surprisingly small labor force, and it is said that no slave was ever bought or sold on Coffin Point, the owners depending solely on those born on the property. In 1813, Coffin records 67 "hands," 16 of whom were too old or too young to be useful. One was a full-time carpenter, three were cattle minders, and one a poultry woman. The house that year required 12 "house negroes," and their rooms were provided in the partitioned-off tabby basement. In addition to planting, Coffin operated a shipyard which did extensive repair work and perhaps even some shipbuilding.

St. Helena was too small for its plantations to be divided, and only one of Coffin's six children remained on the Island. His son Thomas, a Harvard graduate, continued to run the plantation with great success, and Coffin Point eventually included 2,000 acres and 260 slaves. Prominent in Charleston society, Thomas and his wife kept a fine townhouse on the Battery and a summer home in Newport, Rhode Island. When the Union occupied the Island in 1861, the Coffins left, never to return. Thomas' large library was confiscated and shipped North. The following years an abolitionist reported these relics of the remaining furniture: "rosewood tables, sideboards, and wash stands with marble tops, sofas that must have been of the best."

Coffin Point's absent owner died in Charleston in 1863. The year before, his home had become one of the headquarters for the Port Royal Experiment. Newly freed slaves were to be educated and trained so that they could be employed for wages; one of the earliest and best known educators involved in this endeavor was superintendant Edward S. Philbrick. He lived at Coffin Point and bought a portion of the property as well as seven other plantations, which he sold—at a handsome profit—when the war ended.

In 1890, Pennsylvania Senator James Donald Cameron purchased Coffin Point. His wife Elizabeth entertained their close friend Henry Adams there and thus earned a mention in the *Education of Henry Adams*. In 1952, Coffin Point became the home of the high sheriff of the Low Country, Ed McTeer.

Seaside (Fripp)

Built ca. 1800, St. Helena Island, St. Helena's Parish, Beaufort County, National Register.

One of the largest families on the islands, the Fripps appear in colonial records of 1695, and are known to have occupied St. Helena in the early 1700's. Credited simply to this family, the house's builder is unknown, and the date of construction can only be estimated at about 1800.

The original shape of the house was a T, but further additions have obscured this early plan. The house had a curved staircase, woodcarving, and an oak mantel with a painted mural. The Adamesque details are said to resemble closely those of Beaufort's John Mark Verdier House.

In 1863, most of the properties on St. Helena were confiscated and sold by the U. S. Government for failure to pay taxes; Fripp Plantation was one of the few to escape this fate. When he inherited the house from his uncle, Edgar W. Fripp was still a minor, so at the end of the War he was able to reclaim the house and half of his acreage.

Brick Church

Built ca. 1855, St. Helena Island, St. Helena's Parish, Beaufort County.

Following the Revolution, the Baptist denomination became increasingly popular, particularly on St. Helena Island. In the 1830's a small wooden building was erected on what was then John Fripp's Corner Plantation, and in 1855 it was replaced by the present Brick Church.

The slave population preferred this faith to others—at this time there were 156 white and 3,557 black Baptists in the Beaufort area. Even the ample gallery of this building would not hold all these worshippers, and many of the black congregations met locally in makeshift "praise houses."

In 1861, when the church was abandoned by the planters, the freedmen moved their services there, usually with a white minister. From 1862 to 1865 the building housed the Penn School of Ellen Murray and Laura Towne. By the spring of that year Murray was teaching 186 pupils almost single-handedly while Towne attended to health care about the Island.

Laura Towne left the following description of church services:

> Yesterday we attended the Baptist Church deep in the live oaks with their hanging moss. It was a most picturesque sight, the mules tied in the woods, the oddly dressed Negroes crowding in. Inside it was stranger still, the women's turbans, head kerchiefs, or still, small braids—the jetty faces, the men, some in carpet suits, some had on pink calico trousers.

During this time the Brick Church became the religious and social center of St. Helena, and following the War it remained in the hands of the freed slaves.

Rosehill

Built ca. 1860, Colleton River, St. Luke's Parish, Beaufort County, National Register, Open to the Public.

It seems unavoidable that by the end of this survey the words "unique" and "exceptional" should begin to sound stale. Nevertheless, this example of Gothic Revival architecture certainly deserves superlatives. Nowhere before have we seen such a building. The gables of Pine Grove and the icehouse of Bleak Hall barely hint at the grandly asymmetrical and sophisticated combination of Gothic devices of Rosehill.

The layout is cruciform. The steep pitched roof of standing-seam metal is broken by sharp-gabled windows with diamond panes. The siding is board and batten, and the porch roof arches on clustered piers. Each element is designed with great care and consideration for the whole.

The history of the Rosehill site can be traced back to a large grant received by Sir John Colleton. The property remained in the Colleton family until James Kirk bought it from Sir John's great-granddaughter in 1828. Kirk possibly gave the tract to his daughter Carolina, on her marriage to her first cousin, John Kirk. The bridegroom had been educated at South Carolina College and received his M.D. from the University of Pennsylvania in 1834. He was a successful planter and practitioner; his estate was valued at $152,000 in 1860.

Shortly before the War, the Kirks began construction of the Rosehill house. The Yankee invasion came so suddenly that they were caught literally in the midst of a meal. Moving to the relative safety of nearby Grahamville, the owner and his son continued to travel back and forth in an attempt to maintain their estate. Mrs. Kirk died in

Seen through Rosehill's main entrance, a curving stair of teak, walnut, and oak rises to the open gallery above. The chandelier hangs supported from a 54′ domed ceiling. Below the stairs, a curved door of heart pine follows the concave line of the wall.

High roof, steep gables, and asymmetrical form distinguish Dr. Kirk's Gothic Revival mansion.

the village in 1864. After the War her widower managed to hold onto his property, but was unable to complete the house. Various tenants occupied the home but no attempt was made to finish the interior. Finally the Kirk family gave it up in 1928. In 1946, however, Betsy and John Sturgeon, III, purchased and restored in a manner as lavish as Dr. Kirk could have wished.

Although plan books existed for such building, it seems likely that Rosehill was the work of a professional. The job has been traditionally credited to the "French architect Dimmick," who was also said to have designed the Gothic Episcopal Church in nearby Bluffton. Mrs. Iva Welton, Rosehill's present guardian, points out that Dimmick was actually the Kirks' overseer, and that the only architect associated with the Church is E. B. White, who placed a newspaper ad requesting bids for its construction. No firmer link than that has yet been made between that capable Charleston architect and the house.

The Gothic Revival style had been the subject of great debate in the decades before the Kirks began construction. In reaction to the rational classicism of traditional Georgian architecture, the romanticism of England and New England called for a return to the style of peasant cottages from the Middle Ages. Since the beginning of the nineteenth century, most of Kirk's neighbors had been content to build conventional, functional houses, reserving rich embellishments for interiors and furnishings. What they thought of this house at its inception is impossible to say, but in the years after the war, the unfinished mansion was known as Kirk's Folly.

Dr. Kirk's perception of his situation, however, seemed accurate enough. For better or worse he recognized the end of an era, the end of the planter's life as he had known it. Shortly before his death in 1868, he advised his daughter:

> In this climate, a bare subsistence only can be made and a family reared without education and entirely unfit for anything else than the plow handle, the wash tub, or the cook pot. Henceforth, the aristocracy of our country will consist of merchants and professional men.

This was certainly a prophetic vision of the century to follow, but it seems doubtul that even he could have forseen the calamities ahead as bitter Reconstruction gave way to an often destructive self-government, and the boll weevil and Great Depression pushed the South further into grinding poverty. Many of the houses and churches which had survived the War and its aftermath were at last abandoned to decay.

The second half of the twentieth century, however, has seen a reversal of this trend. At perhaps the darkest hour of the mid-1930's, efforts were begun to restore some of the antebellum buildings. The nostalgia of the last generation of plantation-born writers fueled a growing concern for what had been. Families clung to their property when possible, and when not, appreciative buyers were found from outside the South. By hook or crook a legacy was salvaged so that now we can look back at this period and see a grand variety of dwellings, each unique in its own way, and each worthy of protection and admiration.